Art and the City

Artistic practices have long been disturbing the relationships between art and space. They have challenged the boundaries of performer/spectator, of public/ private, introduced intervention and installation, ephemerality and performance, and constantly sought out new modes of distressing expectations about what is construed as art. But when we expand the world in which we look at art, how does this change our understanding of critical artistic practice?

This book presents a global perspective on the relationship between art and the city. International and leading scholars and artists themselves present critical theory and practice of contemporary art as a politicised force. It extends thinking on contemporary arts practices in the urban and political context of protest and social resilience and offers the prism of a 'critical artscape' in which to view the urgent interaction of arts and the urban politic. The global appeal of the book is established through the general topic as well as the specific chapters, which are geographically, socially, politically and professionally varied. Contributing authors come from many different institutional and anti-institutional perspectives from around the world.

This will be valuable reading for those interested in cultural geography, urban geography and urban culture, as well as contemporary art theorists, practitioners and policymakers.

Jason Luger is an urban researcher and Lecturer at the University of California, Berkeley, United States.

Julie Ren is a Fellow in Human Geography at the London School of Economics, United Kingdom.

Routledge Critical Studies in Urbanism and the City

This series offers a forum for cutting-edge and original research that explores different aspects of the city. Titles within this series critically engage with, question and challenge contemporary theory and concepts to extend current debates and pave the way for new critical perspectives on the city. This series explores a range of social, political, economic, cultural and spatial concepts, offering innovative and vibrant contributions, international perspectives and interdisciplinary engagements with the city from across the social sciences and humanities.

For a full list of titles in this series, please visit www.routledge.com/ Routledge Critical Studies in Urbanism and the City/book series/RSCUC

Urban Subversion and the Creative City
Oliver Mould

Mega-Event Mobilities
A Critical Analysis
Edited by Noel B. Salazar, Christiane Timmerman, Johan Wets, Luana Gama Gato and Sarah Van den Broucke

Gentrification as a Global Strategy
Neil Smith and Beyond
Edited by Abel Albet Mas and Núria Benach

Art and the City
Worlding the Discussion through a Critical Artscape
Edited by Jason Luger and Julie Ren

Art and the City

Worlding the Discussion through a
Critical Artscape

Edited by Jason Luger and Julie Ren

Routledge
Taylor & Francis Group

LONDON AND NEW YORK

First published 2017 by Routledge

2 Park Square, Milton Park, Abingdon, Oxfordshire OX14 4RN
711 Third Avenue, New York, NY 10017

Routledge is an imprint of the Taylor & Francis Group, an informa business

First issued in paperback 2018

British Library Cataloguing-in-Publication Data
A catalogue record for this book is available from the British Library

Library of Congress Cataloging-in-Publication Data
Names: Ren, Julie, editor. | Luger, Jason, editor.
Title: Art and the city: worlding the discussion through a critical artscape/edited by Julie Ren and Jason Luger.
Other titles: Art and the city (Routledge (Firm))
Description: New York: Routledge, 2017. |
Series: Routledge studies in urbanism and the city |
Includes bibliographical references and index.
Identifiers: LCCN 2016056813 | ISBN 9781138236219 (hardback) |
ISBN 9781315303031 (ebook)
Subjects: LCSH: Art–Political aspects. | Art and social action.
Classification: LCC N72.P6 A755 2017 | DDC 701/.03–dc23
LC record available at https://lccn.loc.gov/2016056813

ISBN: 978-1-138-23621-9 (hbk)
ISBN: 978-1-138-34643-7 (pbk)

Typeset in Times New Roman
by Deanta Global Publishing Services, Chennai, India

Contents

List of illustrations

Figures

Tables

Contributors

Nabila Alibhai is the founder of inCOMMONS, an organization that develops and invigorates public spaces and builds collective leadership. She is also a founder of limeSHIFT, a company that embeds artists into communities to make them their most creative, productive and connected. She was a recent mid-career fellow in MIT's Special Program for Urban and Regional Studies where she concentrated on models for building community solidarity through public spaces. Prior, she worked for the Aga Khan Development Network, the United Nations and the International Organization for Migration, and has worked on projects in Afghanistan, Pakistan, Kenya, Tanzania, the United States and Switzerland. She has a Master of Public Health from Yale University and is trained in conflict resolution.

Yazmany Arboleda is a Colombian American artist based in New York City. In 2015, he cofounded limeSHIFT, a company dedicated to activating communities through social practice art. limeSHIFT matches artists to companies to co-create inspirational art that aligns organizations on culture, values and brand. An architect by training, Yazmany's practice focuses on creating 'Living Sculptures', people coming together to transform their experience of the world. He lectures internationally on the power of art in public space. In 2013, he was named one of *Good* Magazine's 100 People Making Our World Better. His work is motivated largely by political, cultural and social circumstances. He is also Creative Director of the Brooklyn Cottage, Associate Director of Communications for Artists Striving to End Poverty, and a co-founder of the interdisciplinary performance collective, Shook Ones.

Luisa Bravo is an architectural engineer and an academic scholar, educated in Italy, the United Kingdom and France. Since completing her PhD, with a thesis on contemporary urbanism, at the University of Bologna in Italy (2008), she has been researching, teaching and lecturing in several universities, in Italy, the United Kingdom, Spain, Portugal, Cyprus, Lebanon, the United States, Hong Kong, Taiwan, Japan and Australia. Visiting scholar at University of California Berkeley (USA, 2012), Visiting Assistant Professor at Lebanese American University in Beirut (Lebanon, 2015), she is currently Adjunct Professor at University of Florence in Italy and Adjunct Associate Professor at Queensland University of Technology in Australia. Luisa is charter member and president of City Space Architecture, a non-profit organization based in Italy that performs as multidisciplinary platform for scholars, professionals, artists and citizens engaged in architecture, public space, cities and urbanity. She is Founding Editor and Journal Manager of *The Journal of Public Space*, jointly established by City Space Architecture and the Queensland University of Technology in partnership with UN Habitat, the United Nations Agency on Cities and Human Settlements.

Gretchen Coombs is an academic exploring socially engaged art practices in the United States, the United Kingdom and Australia. She combines the skills from her PhD in social and cultural anthropology with her MA in visual criticism to write essays on contemporary culture. Gretchen is a Lecturer in the School of Design at Queensland University of Technology in Brisbane, Australia, and teaches undergraduate courses such as Culture and Design and Design and Sustainability. She delivers guest lectures on Gender and Design, Art and Social Practice, Design and Ethnography, and runs writing workshops nationally and internationally. She is the Assistant Editor for *Art & the Public Sphere*.

Harriet Hawkins is a Professor in Geography at Royal Holloway, University of London, UK where she co-directs the Centre for GeoHumanities. Her research focuses on the geographies of art work and art worlds, and often proceeds through creative practice-based collaborations with artists and arts organisations.

Marika Hedemyr is a Swedish artist working across choreography and public art, exploring the emotional and political relations between people and places. Her work is known for twisting reality into precise and humorous everyday situations. It has been presented at GIBCA – Gothenburg International Biennale for Contemporary Art, Bibliotheca Alexandrina Egypt, Rawabet Theatre Kairo, and Yokohama Dance Collection, Japan, among others. In 2016 Hedemyr co-edited *Dansbaren – The Mob Without Flash* (Dansbyrån, 2016), a printed object consisting of a book dedicated to critical thinking in the field of dance, choreography and self-organisation, and a tablecloth of topics to create new dialogues. In 2016–2017 she developed *Next To You* – a site-specific multimedia walk for public spaces.

Silvie Jacobi is a London-based cultural industries researcher and urban geographer. Her joint doctoral research at King's College London and Humboldt University Berlin examines the relationship between art schools, location and creative cities. Being trained as a visual artist, Silvie's work contributes to the growing discourse between art and geography. Outside academia she works at a community-run artisan workshop, which is developing into an accredited school. She also engages as independent researcher in projects in the field of regional development and artistic practice in the United Kingdom and Germany.

Liza Kam Wing Man is Assistant Professor at the Department of East Asian Studies at the Georg-August University of Goettingen and lecturer at the

Bauhaus University of Weimar, Germany. She pursued her architectural training in Hong Kong, Singapore, Liverpool, London and Paris before she received her doctorate in architecture (Urban Heritage) from the Bauhaus University Weimar in 2014. Her research interests range from public space and activism, colonial urban heritage and their transformation in terms of symbolism in post-colonial cities in Asia, religious space, housing, and food.

Friederike Landau is a Berlin-based urban sociologist and political theorist, currently working on her dissertation at the Center for Metropolitan Studies. Her research focuses on new forms of artist-led activism and (collaborative) cultural governance. Friederike's project brings together urban sociology, cultural geography, critical policy studies and post-foundational political theory to investigate processes of mobilization and articulation amongst Berlin-based artists. Resulting from this collectivization, Friederike is interested in the a(nta) gonistic governance settings that occur between artist policy stakeholders and the cultural administration. Prior to her dissertation, Friederike worked as strategy consultant, engaging with diverse actors from Berlin's cultural administration, cultural institutions and the independent art scene.

Pei-yi Lu is a Taipei-based curator, researcher, art critic and academic. She has a PhD in Humanity and Cultural Studies (London Consortium), University of London. Her research interests are off-site art, museum studies and curating in theory and practice. A research-based book edited by her, *Contemporary Art Curating in Taiwan (1992–2012)* was nominated 'Art Publication of the Year' in the tenth Award of Art China. Her publications include 'Off-site art in Taiwan, Hong Kong and China' in the Special Issue of *Yishu: Journal of Contemporary Chinese Art* (2010), the edited book *Creating Spaces – Post Alternative Spaces in Asia* (2011) and the book *Off-Site Art Curating* (2011). She was an associate curator of eighth Shenzhen Sculpture Biennale 'We have not participated' (2014), curator of 'Micro Micro Revolution' (2015) for Centre for Chinese Contemporary Art (CFCCA), and co-curator of 'The 10thTaiwan International Video Art Exhibition-Negative Horizon'. She had been teaching in the Taipei National University of the Arts and the Chinese University of Hong Kong. Now, as an assistant professor in the Department of Cultural Creative Industry, National Taipei University of Education and programme leader, MA in Critical and Curatorial Studies of Contemporary Art.

Thierry Maeder is a PhD student in urbanism and spatial planning. He is affiliated with the Institute for Environmental Governance and Territorial Development at the University of Geneva, where he is focusing on the geography of cultural policies. His doctoral research analyses the evolution of artistic public commission and how it articulates with urban planning and the production of public space. Besides his research, he teaches urban planning to Master's degree students. Thierry Maeder has also worked on various mandates for both the public – the Geneva Contemporary Art Fund, in charge of public art policies – and private sector.

Nela Milic is Contextual and Theoretical Studies coordinator and link to Spatial Communication programme in the Design school of LCC, University of the Arts, London. Her research deals with the intersection of time and space, which brought her to many multi-media and arts projects where she dealt with mapping, memory, narrative, archives, protest and participation, all in the urban contexts. Nela is an artist and cultural manager with twenty years' experience, working in the United Kingdom and abroad.

Ben Parry works as an artist, curator and independent researcher at the intersection of art, activism and urban space. He is Programme Leader of the MA in Curatorial Practice at Bath School of Art and Design. He is editor of the book Cultural Hijack: Rethinking Intervention (Liverpool University Press, 2012). He curates the project Cultural Hijack, an international survey exhibition, set of symposia and live-programme of urban intervention and art activism, at Architectural Association, London, 2013; Archip, Institute of Architecture, Prague, 2017, and Delhi, 2018. Working under the International Peripatetic Sculptors Society (IPSS), he is regularly invited to conduct action-based workshops fusing the Situationist tradition of dérive with the creation of spontaneous interventions.

Luca Pattaroni is a sociologist and senior scientist at the Laboratory of Urban Sociology of the Swiss Federal Institute of Technology (EPFL) where he leads the research group 'City, Habitat and Collective Action'. Focusing on margins and social movements (squats, artistic counterculture, public events, urban struggles), his work is concerned with the making of the common in contemporary capitalist cities through a critical study of urban policies and assemblage. His work has been published in, among others, the *International Journal of Urban and Regional Research*, *Urban Studies*, *Swiss Journal of Sociology*, *Sociologies*. In 2009, he received the APERAU prize for the best scientific article on urban planning (in collaboration with Vincent Kaufmann, G. Pflieger and Ch. Jemelin). Along with his academic commitment, Luca Pattaroni works frequently as an expert with urban planners, architects and public agencies and is president of the Geneva-based cultural cooperative *Ressources Urbaines*.

Mischa Piraud is a PhD student at the Urban Sociology Laboratory (LaSUR) of the Swiss Federal Institute of Technology (EPFL). His research interests revolve around the metamorphoses of capitalism and politics of art in the contemporary city. He published *De la Différence Urbaine* (Métis Presses, 2013) with Elena Cogato-Lanza, Luca Pattaroni and Barbara Tirone.

Brian Sonia-Wallace is a Los Angeles-based poet, performance artist and satirist, who challenges notions of public intellectualism and explores themes of connectivity and spatiality. He was nominated for a 2015 Doris Duke Impact Award in part for his 'RENT Poet' project, and has received artistic residencies from Amtrak (2016), City of Los Angeles (2016), National Parks System (2015), Dollar Shave Club (2014), and the Shuar Nation of Ecuador (2012). He creates work ranging from poetry performance to clowning to bike ride, including *Nevermore* (2016), *Quick Draw Poetry Cabaret* (2016), *White Man,*

Trying to Help (2015), *Bike Odyssey LA* (2014), and more. His first book, *I Sold These Poems, Now I Want Them Back*, was published by Yak Press in 2016. Brian holds an MA in Sustainable Development from the University of St Andrews, Scotland.

Jennifer Teo is a Singapore-based artist/curator who co-founded Post-Museum in 2007.

Woon Tien Wei is a Singapore-based artist/curator who co-founded Post-Museum in 2007.

Phoebe Wong is a Hong Kong-based culture worker with a special interest in contemporary art, design and visual media. She was Head of Research at the Asia Art Archive before becoming an independent researcher and writer in 2012. She is currently a member of the International Association of Art Critics, Hong Kong. Wong is also a member of the Community Museum Project, a research and curatorial collective dedicated to revaluating indigenous creativity and the under-represented histories and practices of the everyday.

Sampson Wong is a Hong Kong-based urban geographer artist and independent curator. He is currently a lecturer of the Department of Liberal Arts at the Hong Kong Academy for Performing Arts. His research focuses on urban political ecology, urban interventions, public space and Hong Kong studies. He co-founded the arts collective Hong Kong Urban Laboratory and has directed the Umbrella Movement Visual Archive.

Acknowledgements

The editors would like to acknowledge the institutional support of King's College London and the National University of Singapore, Departments of Geography, that provided, among other things, conference funding to attend AAG 2014 and 2015 for Jason Luger. It was out of the AAG sessions that the ideas and collaboration framework for this book were generated. Jason is also grateful to Professor Loretta Lees for her doctoral supervision and for helping to organize the King's College London/National University of Singapore departmental partnership, which provided research funding. Julie would like to thank Ilse Helbrecht for her encouragement, and invaluable guidance in navigating the landscape of urban geography. Further institutional support was provided by the Humboldt University Berlin, City University of Hong Kong and the London School of Economics, all of which helped Julie Ren make the work on this project possible from 2014 to 2016. Thank you to the contributing authors for all the hard work and dedication to this project.

Foreword

Arts works: Art worlds – scales, spaces, scapes

Harriet Hawkins

Worldings

What does it mean to talk of a global art theory, to think in the terms of this volume and other texts, of a 'worlding' of the art world? Like all 'worldings', those of the art world encompass many different senses of the word. It might mean to assert the 'global' nature of the art world, to sensitise us to those theories and practices of art that originate from beyond the territories of the 'Western' art world. This is a 'worlding' that takes seriously those overlooked and under-engaged spaces and practices of art, resisting the normative situation of the Western art world's 'other' as a tokenistic case-study and seeking rather to see it as a site for the generation of theory and practice. It is a worlding that demands we move beyond the stereotypes through which we tend to understand cities of the global south (McCann *et al.*, 2013). To speak of 'worldings' is also to evoke the globe-trotting nature of much Western contemporary art. A sort of false 'worlding' often dismissed as its 'one-place-after-another' tendency, this is not a lack of 'geography' per say, but rather the evolution of a geography that is perhaps less attuned to site, less engaged with location (Kwon, 1997).

But to muse on 'worldings' is also to think in other terms too. Terms that while not initially as directly tied to the global are, I think, crucial when it comes to thinking through what it means to think of, or for, an art theory understood in the context of ideas of 'globe' and 'world'. It is three of these I want to explore here. First, I want to consider scale. For ideas of 'world' and 'globe' conjure imaginaries that bring with them a particular rhetorical gravity as well as a sense of agency that are worthy of further critical consideration. Secondly, space. For, when thinking about the worldings of our global artscapes, we should remind ourselves of discussions of the relationship between art and space. These are discussions that foreground the always co-constituted nature of art and space, whether or not that finds presence in the art-work itself. Finally, 'scapes'. It would be remiss to explore artscapes without taking the chance to reflect on the rich intellectual inheritance that the idea of 'scapes' brings. Drawing on the cultural geographies of landscape, I want to reflect on politics, embodiment and the 'work' of artscapes.

Scale

World- globe-ing, to speak of globes and worlds is to speak of different, but intersecting, entities. It is also, perhaps inevitably, to assert some form of scalar imaginary. In the midst of a conversation riven with the expansive scales of globes and worlds, it seems important to pause and reflect on such 'up-scalings', what they put at stake, and what they might offer.

Geographers have gone up and down and back and forth with the global and the local, with the latter being seen as at worst as a victim and at best as a source of 'bottom-up' empowerment. Taking lessons from the debates within geography about scale would suggest a need to denaturalise the global as the ultimate goal, and instead to think in more nuanced ways about its relationship to the local (Marston *et al.*, 2005). This might be about reflecting on the local as less a victim, an overlooked and ineffective underdog, and instead to seek forms of understanding and empowerment that work from the bottom up. Indeed, as McCann *et al.* (2013) note, such 'worlding' practices should not be those focused on governing and transnational elites, and the reproduction of accepted models of urbanism, but rather might embrace new forms of political process and explore how ordinary people, in this case artists, make their place in the world.

This is to appreciate that the global is not necessarily the apex of aspiration, instead an appreciation of local specificity in the face of the 'global' might be appropriate. To sit with the local is to appreciate difference, to refine and develop specific senses of the production and consumption of art work and its sociabilities. It is to recognise that an appreciation of the local is not to suggest that the global, and an appreciation of global differences is not vital, but is to ensure that the global is not somehow seen as ultimately better, as more worthwhile, as more important.

Space

Art and space are inseparable, and not just for geographers, whether thinking of the spaces of art's production and consumption or those through which it circulates, or whether concerned with the site of the work – the space of the picture frame or installation and the encounters that are created. To think of art is to think of space. But yet art is not just *in* space, it is also *of* space, and importantly space is *of* art. In other words, when we think about the relationship between art and space we should to be aware of how space is not a mute, already existing background across which art is produced and consumed, rather the making of art and the making of space are entwined.

If one aspect of the 'worlding' of art is to explore how it is that the worlds in which art is made might shape it, whether thinking about site specificity or about the myriad roles of place in art's formation (Kwon, 1997; Dacosta Kaufmann, 2004). Then a second, and perhaps the worlding of art works that I am most interested in here, are the worlds art works make. By this I mean not only, the spaces of studios, galleries, art networks and so on, but rather the ways that art goes to work in the world, the ways that the work might shape subjects, develop communities, make worlds (Hawkins, 2014).

Of course, the 'spaces' that the art is making and made by, take many forms, depending on the forms of artscape being talked about (Kwon 1997). Whether this be the social spaces of socially engaged art, where human and non-human assemblages both play a role in producing the work, but are also produced by the work (Thompson 2012). Or, these might be spaces of embodied experience found in phenomenologically engaged sculptural or installation practice, or they might be the spaces of encounters in cosmopolitan or not-so-cosmopolitan places (Kester 2013; Kwon, 1997).

Scapes

To think of art worlds is to conjure up networks of studios, galleries, artists, curators, critics and agents. It is to summon up those creative economy and sociological approaches to creativity that enable the production of cartographies of creative production that identify and explore actors (human and non-human) and their relations in the creative process. Yet, to talk of artscapes offers us something else, amongst other things it directs us toward that binding together of politics and aesthetics that sits at the heart of the geographies of landscape.

Landscape or artscape, to speak of 'scapes' is to do more than to speak of spaces or worlds. It is to engage with specific questions of the aesthetics of experience but also to evoke a deep legacy of politics and poetics. It is to find tensions between looking and feeling, between politics and aesthetics, between pleasure and labour (Daniels, 1993; Wylie, 2007).

By aesthetics here I mean both those formal codifications of the iconographies of the picturesque and the sublime, but also the more generalised formulation of aesthetics as an attunement to sensory experiences. To think of 'scapes' enables us to appreciate these aesthetic dimensions of art worlds. It is to foreground our embodied experience of situated assemblages of humans and non-humans. It is also, however, to appreciate how these aesthetic experiences take place not only through local adaptations and alterations but also through global circulations, as artists and art works travel through art festivals, as well as mediated by the internet.

In their formulation within landscape, 'scapes' bind aesthetics with politics. For scapes are socially produced spaces, they are sites where dominant cultures will assert themselves and where the art most often celebrated and supported is that which is produced by, and which reproduces that dominant culture. As such artscapes, like landscapes, should be considered, at least in part, to be under the control of a range of forces, from those of government, to those of arts governance, arts markets and institutions as well as theorists and high-profile individual practitioners.

Yet artscapes, like landscapes, might offer us other opportunities, other than being appreciated as the product of the powerful (Mitchell, 1996). For just like landscapes, we might witness calls to appreciate the overlooked labour that can go into their making. We might also vision an assertion of the aesthetics and life-worlds of the less dominant cultures that scapes, land or art, enable us to experience. It seems then, that a crucial lesson we might learn from landscapes is the importance of enabling us to presence those less elite forms of labour, experience and local life worlds that constitute artscapes, wherever in the world these may be.

Art works – art worlds

Worlding art works, worlding art theory, worlding art worlds, these are slippery and multi-faceted endeavours. Endeavours that at their heart speak to the ways that artists make works, the ways that art works make artists, the ways that audiences (intentional or otherwise) consume work, and the ways that works consume audiences, the ways that art works make worlds and that worlds make art works. There is 'worlding' work to be done in art theory to be sure, but whatever forms that takes, we should ensure that the worlding we never overlook in our seeking of global theories of art found in the myriad world-makings that art, wherever around the globe it is produced and consumed, brings about.

References

Dacosta Kaufmann, T. (2004) *Towards a Geography of Art*. Chicago, IL: University of Chicago Press.

Daniels, S. (1993) *Fields of Vision: Landscape Imagery and National Identity in England and the United States*. Princeton, NJ: Princeton University Press.

Hawkins, H. (2014) *Creative Geographies: Geography, Visual Art and the Making of Worlds*. London: Routledge.

Kester, G. (2013) *Conversation Pieces: Community and Communication in Modern Art*. Oakland, CA: University of California Press.

Kwon, W. (1997) *One Place After Another*. Boston, MA: MIT Press.

Marston, S. A., Jones, J. P. III and Woodward, K. (2005) 'Human Geography without Scale'. *Transactions of the Institute of British Geographers*, 30: 416–432.

McCann, E., Roy, A. and Ward, K. (2013) 'Assembling/Worlding Cities'. *Urban Geography*, 34: 581–589.

Mitchell, D. (1996) *The Lie of the Land: Migrant Workers and the Californian Landscape*. Minneapolis, MI: University of Minnesota Press.

Thompson, N. (2012) *Living as Form: Socially Engaged Art from 1991-2011*. Boston, MA: MIT Press.

Wylie, J. (2007) *Landscape: Critical Ideas*. London: Routledge.

Introduction

Disruptions of a critical artscape

Julie Ren

> Maybe this is, or could be, one of the potentials of landscape as a provocation.
> —Doreen Massey (2006: 43)

Artistic practices have long been disturbing the relationships between art and space. They have challenged the boundaries of performer/spectator, of public/ private, introduced intervention and installation, ephemerality and performance, and constantly sought out new modes of distressing expectations about what is construed as art, where it might be found, how it is created and for what purposes. If all art is in part about the world in which it was made, this certainly begs the question about what kind of world produces art, and what kind of impact art is seeking to make in this world. The question of the object of critique in art is partnered with the question of the contextual origins of criticality itself. In other words, when we expand the world in which we look at art, how does this change our understanding of critical artistic practice?

This volume brings together perspectives on critical artistic practices that encourage the 'worlding' of the concept of the 'artscape'. Both the notion of worlding and the concept of artscape share the quality of seeking disruption rather than stability. Rather than offering a categorical certainty for typologies of art practice, the contributions in this volume defy easy interpretations. Thus, in multiple ways, the question of disruption seems at the heart of much of the present volume, disrupting expectations, interpretations and framings of both art practice and the places in which they take place. In lieu of a comprehensive theoretical framing, this introduction will first discuss these two concepts and the existing literature as the formative rationale for this volume before presenting some notable findings from the individual chapters and the outline for the volume as a whole.

Favouring disruption over cohesion in terms of the framing of this volume is a choice as well as a reflection on the contributions. As a choice, it is about the book as a project: how to design a volume connected to these topics? What kind of contributors should be included? The contributors include activists, artists, curators, researchers and many who would identify with multiple of these roles, and other roles in addition. They come from a diversity of geographic and disciplinary backgrounds including anthropology, architecture, art history, geography, planning, sociology and more. Their contributions include art works of their own and of others, presented with varying analyses of how a variety of art practices, and practices that facilitate art functioned as critique or intervention in their different contexts. Choosing breadth over stringency implied the volume had the potential to disrupt some presumed understandings about critical art practice itself.

Focusing on disruption is also a reflection in terms of recognizing the various concepts, practices, references and goals of the artistic practices themselves. They are disruptive in their lack of cohesion, as well as disruptive in their interpretation. They elude a coherent theoretical framing and demand some kind of bridging concept that might capture the quality of unsettling rather than commitment to dominant canons of interpretation. In evaluating the book proposal for the present volume, an anonymous reviewer noted that there seemed to be 'little concern for precedents or the histories of critical art practices from, say, Situationism onwards'. Reflecting on these chapters, however, the issue of precedent and history demands an active disentangling from the rather myopic heartlands of critical art practice. Interpreting the work of 'Monday Mornings' (Kabul, Afghanistan, 2014, depicted on the cover) by contributor Yazmany Arboleda, where balloons were distributed through a city the day after a suicide attack, in terms of the work of Guy Debord feels reductive, at best. Moreover, it is unclear whether there is any shared history, or how many distinct histories might be useful in understanding the critical art practices presented in this volume.

While the challenge of framing the histories and contexts of these works is left to the individual contributors, the focus on disruption deserves some more discussion here.

'An urge to unsettle'

The notion of 'worlding' has been adopted and developed by a number of urban scholars as a means to disrupt how cities are represented in urban studies, and how they are empirically codified (McCann, Roy and Ward 2013; Roy and Ong 2011). Worlding is a practice of intervention, which attempts to 'establish or break established horizons of urban standards in and beyond a particular city' (Ong 2011: 4). In doing so, it serves as a critique of the tendency in urban studies to isolate cities or categories of cities from one another.

Rooted in fears of developmentalism or established modes of comparative research, the strictures of relating cities across presumed differences has resulted in an uneven field of urban knowledge (Robinson 2011; McFarlane 2008). That some cities serve as sites of theory, and others as empirical variations results in 'a case study-ism that precludes alternative grounded theoretical insights' (Ren 2015: 340). Thus, the notion of worlding serves a purpose within urban studies. Worlding is an active process of challenging representation of difference between cities to expand the sources of urban concepts.

Worlding is not, however, purely in the selection of contributions that traverse the map. Ranging from Nairobi and Geneva to Taipei and the US–Mexico border, the sites contained in this volume certainly represent a diverse range of urban spaces. Yet site selection is insufficient in bringing these places out from the 'urban shadows' of urban theory (McFarlane 2008). Attempting to focus on artistic practice in the organization of the volume rather than geographies of difference is a part of the goal to relate these places to one another. Rather than dividing these chapters according to global north/south or east/west, the organisation of these chapters around various practices is intended to disrupt the codification of presumed difference connected to the city. In other words, through bringing these practices together, the goal is to disrupt the assumption that Kabul and Paris are incommensurate sites of comparison.

Worlding is not a scheme of instituting geographic quotas ensuring regional diversity, but the bringing together of these practices in a way that does not predetermine their positions. It serves to trouble distinctions of global or world cities (Roy and Ong 2011) through the focus on critical art practice.

As part of this practice, it seemed that some notion of artscape as a bridging concept would additionally be useful in attempting to theorize what kind of spatiality is ensconced in these works. While the term seems to exist as an implicit idea about cultural landscape (see e.g. Harris 2013), this volume offers the opportunity to be more explicit about what the concept of an artscape could offer. Rather than a static or objective setting, however, it seemed much more useful to borrow from the enormous literature on landscape for some inspiration. At its core, the artscape disrupts the idea of stability, or, like landscape, resists being reduced to something that is 'timeless, fixed and static' (Hawkins 2012: 63, citing Housefield, 2007; Matless and Revill, 1995; Yusoff and Gabrys, 2006).

Conceiving of the artscape as the conceptual space of expansion and intervention, of provocation and emergence also takes inspiration from Doreen Massey's rumination about 'the potentials of landscape as a provocation' (2006: 43, epigraph above). In her reflective text about landscape, Massey recognizes how her writings on place have exclusively focused on place-making as a human/social activity (2006: 36) and acknowledges the relative absent discussion of physical space. She argues, however, that rather than adopting landscape or terrain as a stable representation of nature to be taken into account as some kind of backdrop for social actors, landscape itself is constantly in a mode of being made, a terrain characterized by its potential. Similarly, conceiving of the artscape like this landscape rich in 'potentials' provides a framework for understanding how the artscape might also serve as a provocation.

Building on a longer trajectory of the conceptual development of landscape, including Burkhardt and Cosgrove, Matthew Gandy usefully offers landscape as a means to promote a 'wider reflection on the meaning of spatial ambiguity, complexity and multiplicity' (2016: 433). Reflecting on the meaning of multiplicity by contrasting the contributions on art activism in Hong Kong provides a useful example of this. It is evident in the issues around representation when focusing on subverting colonial expectations (Kam), or aesthetic congruence (Lu and Wong), or through participatory forms of crowd creation (Wong).

For Massey and Gandy, among many others, landscape is a concept that is both dynamically being made and analytically useful in its disruptive possibilities. While encountering or intersecting with materiality, landscape is not simply a backdrop or terrain. These elements help in thinking about the conceptual usefulness of an artscape.

An analytical concept that does not dictate singularity or coherence, but represents an experience of 'encounter between preconceptions and material elements' (Gandy 2016: 434). Like the conceptual troubling of 'land' as a natural backdrop (see also Scott and Swenson 2015), there is nothing natural or given about the spatiality of art, and its master narrative certainly requires some interrogation. Yet, there remain entrenched codifications of art history that often assume that the heartlands of critical artistic practice remain comfortably situated in New York and Paris, as suggested by our anonymous reviewer, noted earlier in the chapter. Indeed, the reference points for what constitutes 'contemporary art' as opposed to 'world art' remains a debate steeped in US/Euro-centrism (see e.g. Elkins 2007).

This volume demands a flexible concept like the artscape to accommodate the impossibility of a coherent set of reference points, histories or modes of evaluating what exactly constitutes art, or critical art, or in whose terms these various works can be interpreted. But there is a point of unity amongst these widely divergent works. Borrowing from Massey's discussion of landscape, the artscape presented here serves as a conceptual framing characterized by an 'urge to unsettle' (Massey 2006: 43). The artscape is a concept that clearly also takes inspiration from Arjun Appadurai's gesture with his '-scapes'.[1] He schematizes different scapes to help in understanding the fluidity of global cultural flows, as well as the importance of perspective in this (Appadurai 1990). The artscape emerges then as a conceptual tool characterized by this unsettling fluidity.

More specifically, the artscape is encapsulated in the shared importance of criticality. Though the object of critique is widely divergent, for example with political protest, or as explicit modes countering commodification by rejecting the art market, or through subversive representations, the practices hold a critical position close to their heart. But beyond the object of critique in the art work, there is also a critical element in terms of the interpretation of the work within various institutional, political, competitive settings. This is most evident when considering the art practices that are promiscuous in their dealings with institutional formality, especially in regards to funding. The modes of criticality, however, are divergent and at times contradictory. A dichotomous politics of left or right feels anachronistic, if not completely out of place. (For a more nuanced discussion about the political, see Maeder *et al.* in this volume.)

As a whole, the perspectives on art practices in this volume hold in common a perspective that stands in stark contrast to the overwhelming proportion of literature about art and the city preoccupied with its utility. The research in this area largely focuses on an instrumental view of art for urban regeneration, planning or development. The function of art is seen in terms of transforming the image of the city (Grierson and Sharp 2013), making global cities (Chang 2002; Kong 2012; Kong, Chia-oh and Tsu-lung 2015) or as a part of aspiration urbanism (Wang, Oakes and Yang 2015), which could all also be interpreted as a largely congruous mode of instrumentalizing art. More specifically, it is described in its uses for processes of land use reform or property valorisation (Ley 2003; Scott and Swenson 2015) that are seen at the heart of various processes of redevelopment, regeneration, gentrification (see e.g. Cole 1987; Cameron and Coaffee 2005; Lloyd 2006;

Markusen 2006; Wojan, Dayton and McGranahan 2007; Markusen *et al.* 2008). The decades-long debates around public art range from the aesthetic discussion of enhancing public space and regeneration (Pinder 2008), to issues of funding, inclusion and representation (Miles 1997; Sharp, Pollock and Paddison 2005).

While concerned with impact, resonance and relevance, the perspectives on art and the city in this volume are largely uninterested in the instrumentalization of art for the city. This should be hardly surprising considering how few urban specialists are included in this volume. Yet, how is it possible to read these contributions alongside each other? While the instrumental view of art and the city offers some cohesion, the artscape seems at its base to be deconstructive. Is the artscape at all useful as a bridging concept to bring these works into some shared context of disruptive unsettling?

To consider the possibility of comparing what anti-establishment might mean for an artist in Los Angeles contrasted with Singapore, or the consequences of participatory practice in Hong Kong during the Umbrella Movement for crowd-sourced creations in comparison with the impact of participatory practices in public art funding in Geneva for right-wing activists, the issue of context demands attention. Situating each city and artist within their particularistic, historical origins seems inappropriate considering the connected nature of these contributions. Indeed, it is perhaps through a closer look at some of the emergent themes that it is possible to evaluate the relevance of the artscape concept.

Among the many themes that are recurrent throughout these contributions, two issues could be featured as emerging out of the terrain of a critical artscape: duress and difference.

The issue of duress is demonstrably present, especially as it results from violence, authoritarian encroachment, states of emergency and the concomitant precariousness. The duress is evident in Arboleda and Alibhai's work in Kabul just following a violent suicide bombing (Chapter 1), in Lu and Wong's chapter, which elicits the spectre of Chinese state encroachment in Taipei and Hong Kong (Chapter 3), or in Parry's discussion about the state of emergency during the United Nations (UN) climate summit in Paris following terrorist attacks (Chapter 9). The underlying threats transform the significance of a balloon in 'Monday Mornings', and, as Parry describes, they 'set off an alarming performance of politics, power and ideology' that would render 'fear a constitutive element of everyday life' (see p. 129). Duress is not an exceptional situation, but a normalized state of fear and art interventions that take over public space as in these works in Kabul, Taipei, Hong Kong and Paris, serve as a disruption of the securitization. Just as the materiality of the landscape is mutable, the material aspects and even art practices are rendered disruptive in large part set against the state of duress.

Volatility as a source of art practice might lead to some troubling implications. While it might suggest that these works are only significant given the duress, the relationship between the art works and these settings of fear is to attempt to transform or subvert them. This happens through the use of public space in ways that are emotionally expressive, through the large-scale works that are coordinated in their use of colour like 'The Red Lines' in Paris in which 15,000 people in red or carrying red objects moved through the streets as a statement about the red lines protestors did not want climate negotiators to cross. Similarly, the 'Colour in Faith' project used colour as a statement about unity when it painted different houses of worship yellow in Nairobi. Against a backdrop of volatility and under duress, these coherent images emerge as a shared practice of disruption. These practices engage with a number of issues specific to a space-time, but they seem connected in the terms of their disruption. Though the goals and issues vary, and certainly the nature of the duress is difficult to equivocate, the emergent practices share certain material and conceptual compatibilities that support the concept of artscape.

In important ways, the participants in these large-scale public art interventions are different and represent uneven issues of access and positionality. But for the period of these happenings, these art works have a shared function of disrupting the normalized everyday state of securitization, control and fear. In doing so, it helps to emphasize the disruptive characterization of the concept of artscape.

A second issue deals with the discomfort of difference. Several art practitioners struggle with ideas about 'the community' they work in, which is separate from them but also something they seek to engage. The issue of seeking to respond to 'community' needs whilst also serving to construct micro-utopias for artistic works is highlighted in the history of 'Post-Museum' in Singapore (Chapter 2), located in a former brothel house. The juxtaposition of addressing the 'existing communities which were there' while also directly displacing former residents of the building is a discomforting position that recurs in several chapters. It comes up in the discussion of space in London, when artists engage simultaneously in 'appropriating space' while also engaging the local community (see p. 87). It is revealed in the personal reflections of Luisa Bravo when she positions her gaze, 'From a Western perspective, especially for an Italian used to walking, living in Beirut hurts, because the Arab everyday urbanism hurts' before proceeding to list the chaos, traffic, streets, dust, garbage, noise and lack of amenities that make it difficult for her to walk around (Chapter 14, p. 209).

While it is tempting to think of the disruptions that the artscape offers as somehow uniformly progressive in various critically engaged ways, the artscape is also subject to the position that art, artists and art spaces often inhabit in the city. It is a position that empowers them to highlight the blight of certain neighbourhoods or social groups in Singapore, or to document the ways that art activism in Beirut addresses the commodification of public space. But artists also inhabit a position that forces disruptions that may imply displacement, and impose interpretations that relapse to forms of cultural analysis rejected for their imperial suppositions (Said 1993). The artscape serves as a flexible concept, characterized by its disruptive quality and its critical intentions, but it is also subject to a variety of actors, their perspectives and positions.

Still, many of the works in the volume, especially those directly written by artists, offer a reflexive analysis on the assumptions about difference. The work of Ghana ThinkTank focuses explicitly on this issue of constructed difference (Chapter 4). It is a project run by a collective of artists that seek to reverse assumptions about the sources of expertise that delineate the first/third world construction. But even while reversing the assumptions about development in 'developing the first world', they find a binary distinction difficult to maintain. The border space and the mobile serve to disrupt clear boundaries, as do the differences inherent in the 'first world'. Dealing with these layers of difference makes Ghana ThinkTank a fascinating art work, but also challenging in determining what exactly it is disrupting.

Disruption, duress and difference are all relevant elements of this artscape, itself a concept that attempts to bridge art and the city. Worlding the contours of this artscape, issues emerge that bring places somehow in closer proximity with one another. The practices and experiences throughout this volume resist easy framing, though a flexible concept like the artscape might help to think about the different ways 'an urge to unsettle' might be manifested, and how these works can be interpreted.

Book structure

The volume is divided into three sections to further suggest some possible ways to connect the practices to one another:

1   alternative imaginations
2   transformative processes
3   public(s), participation and representation.

In the first section on alternative visions and representations, the contributions present alternative geographic imaginations (Gregory 1995). The need for alternatives originate from the limitations set by a variety of factors including threats of violence, authoritarian states or entrenched representations that artists seek to subvert.

In Chapter 1, Yazmany Arboleda and Nabila Alibhai present large-scale public art interventions in Kabul, Johannesburg and Nairobi from 2013–2015. In three separate works, the impact of colour is particularly striking set the day after deadly violence, in post-apartheid era dilapidation, or as a means to cohesively celebrate inter-faith tolerance. The works illustrate the possible functions and methods of art under duress, and the chapter offers some reflection on the enabling conditions for implementing them, as well as the resonance of these types of works in the media. In terms of alternative imaginations, the simultaneous use of colour and participatory modes of implementation, serves to disrupt the everyday experience of fear in these public spaces.

In Chapter 2, Jennifer Teo and Woon Tien Wei present the Post-Museum in Singapore, which highlights the role of alternative art spaces within an authoritarian context and in a context of restricted freedom of expression. The Post-Museum sought to 'find ways to create micro-utopias', to imagine alternative worlds that participants would desire, and to give space to works that could not find other outlets or venues (see p. 31). Implicitly, they offer a critique of the dominant institutional art spaces and practices, such as state-led art institutions in Singapore, where experimentation and engaging with social and political issues is often discouraged.

Chapter 3 aims to compare the role of aesthetics in the Sunflower Movement in Taipei and the Umbrella Movement in Hong Kong. Pei-Yi Lu and Phoebe Wong draw on the aesthetic and collective practices within these movements to argue that they could be interpreted as 'a war of visibility' (see p. 55). These movements present an expansion of public art and its makers, which contribute to new imaginations about the societies in which they are taking. These imaginations stand in contrast to the perception of Chinese encroachment in the region, the sense of inevitability and loss that the movements seek to counter.

Chapter 4 takes the question of subversion beyond specific urban context, and towards challenging the idea of development by reversing the assumed positions of expert knowledge. The 'reversal' considers the possibility of an alternative imagination about development, and the role of outside expertise or intervention. Gretchen Coombs conducted interviews with the members of Ghana ThinkTank for her analysis of their works at the Mexico–US border and in Corona, Queens, in New York. These works highlight a number of reflective turns about assumptions of difference that span not only spatial, social, but also temporal (inter-generational) scales.

Section two of this volume is comprised of artistic practices connected to transformative processes that deal specifically with funding and policy. They push the boundaries of artists' roles, especially as they interact in novel constellations with various institutions, and institutional processes of decision-making about resources for the arts. In certain ways, these practises all share a component of co-optation, or the fear of co-optation. In part, this seems to be a result of the artists taking on roles like government lobbying or commercial formats like advertising. Still, they manage in the final reflection to appropriate these forums for their own interests, or to offer subversive analyses distancing themselves from the co-optation.

Chapter 5 presents an artists' coalition in Berlin, that effectively lobbied for public funding to go towards independent art spaces. In tandem with a wider network of actors, Friederike Landau presents the ways in which they ensured the funding was institutionalized within the Berlin city government, effectively serving as policy-makers. The significance of the work of this coalition is not only in the end result of the budget, but also in the way that they changed the processes of consultation and policy-making.

Moving to London in Chapter 6, Silvie Jacobi presents the establishment of alternative art schools in the context of urban regeneration as well as in the pursuit of pedagogic independence. These art schools serve as a reflection on both the competitive urban space challenge for artists in London, as well as the art education system from which art professors sought more independence. By instituting these programs outside the official higher education system, but with professors that are affiliated with accredited art schools, these initiatives transform an entrenched system otherwise resistant to reform.

In Chapter 7, the poet and artist Brian Sonia-Wallace reflects on his own work as a 'RENT poet' in Los Angeles, and the ways that he engenders the process in which 'the anti-establishment becomes the thing most sought by the establishment' (see p. 100). In a careful analysis of the relationship between his artistic practice and the various representations of 'the establishment', including corporate events and museum institutions, the chapter offers a self-aware artist's perspective on the artist economy.

Chapter 8 looks at the way that artistic activism in Hong Kong served as both a disguise for more contentious intentions and a means to attract a broader audience. Liza Kam Wing Man connects the representation of polite Hong Kong to a period of de-colonization, and suggests the subversive implications of art activism as a disguise for political activism in this particular setting. That artistic practice can provide a cover for more contentious political positions is certainly provocative and considers the strategic possibilities for artistic practice.

Ben Parry includes his experience being directly involved with the Climate Games during the UN Climate Change Conference in Paris in 2015, and considers the nature of resistance in a politically charged moment. Chapter 9 is set against the backdrop of a state of emergency in Paris following violent terror attacks in the city, and an analysis of historical allusions that the securitization evoked. It explores some of the connections between the security state, the role of emotion, the 'Brandalism' that was popular with the media and the substantive goals of climate change activism.

The last section of this volume deals more directly with questions of representation and participation, particularly in engaging with a broader idea of the art public. In different ways, these chapters consider the impact of public involvement in the shaping of public space.

Chapter 10 begins with a theoretical discussion about the 'subversive dimension', particularly insofar as it is attributable to public art. Following their analysis of the politics of art from a theoretical standpoint, Thierry Maeder, Mischa Piroud and Luca Pattaroni draw on some examples of public art funding from Geneva to help illustrate the difficulty in drawing clear lines around what constitutes subversive, or the possibility of subversion itself.

Continuing with the question of public space, Chapter 11 looks at the restructuring and branding of Gothenburg from an artist's perspective. Alongside Marika Hedemyr's interpretation of the cannibalisation of the city, she describes her long-term art project The Event Series as a strategy in which more of the complexity of this urban development strategy and its effects can be revealed. Taking place as a kind of tour through the city, it guides an audience through the urban space with the artist's critical narration.

Another artist's contribution is from Nela Milic, in which she describes her residency in a particular area of London that was undergoing dramatic changes. Chapter 12 deals with the question of neighbourhood and social engagement, depicting participatory art as a means to engage with complexity, to mobilize shared interests and also for teaching. The participatory nature of the project is considered in its complexity, with regards to expectations, positions and the implications of the artist's own embeddedness.

Sampson Wong also focuses on participatory art production in Chapter 13, when he discusses crowd creations during the Umbrella Movement in Hong Kong.

Rather than the iconic, professionalized art interventions, the crowd creations were characterized by spontaneity and they were largely decentralized. These crowd-sourced creations served multiple functions for the broad audience of the movement (exogenous), as well as the producers of the art (endogenous). The multiplicity of these functions and audiences serve furthermore as a reflection about the nature of the movement itself.

Finally, in Chapter 14, Luisa Bravo offers a personal narrative of her time in Beirut, situating Beirut within its geopolitical context before exploring the way that activism connected to the city is connected to the Arab cyberspace and the changing idea of public space in general. It highlights how images in the form of photography, graffiti and mural art agitated against the privatisation of public space in the city, and, while ephemeral in the city, gained a permanency through their documentation and subsequently being posted online.

The careful reader will also find multiple connections across the sections that are worth further discussion. This includes some things raised in this introduction, including the role of online archive to extend the temporality of the art work and to expand the public realm of the urban space (Lu and Wong, Bravo), or the impact of cohesive aesthetic choices and scale for media attention like with colour (Arboleda and Alibhai), objects (Wong), or format (Parry). Another volume could certainly spend time addressing the embodiment of the artist in urban space (Hedemyr, Milic, Sonia-Wallace) and reflect on the voice of the artist when writing about their own work – contrasting these reflections with a collaborative text (Arboleda and Alibhai, Teo and Wei) or when it is written by a researcher rather than the artists (Bravo, Coombs). Due to the constraints of the present volume, this further analysis will be left to the reader to pursue.

Indeed, the contributions to this volume encompass an enormously broad body of work, which should incite the reader to consider the many connections and contradictions revealed about this relationship between art and the city. 'Worlding the critical artscape' serves here as an introduction to begin shifting our desire from the definitive, final explanation towards considering the many multiple ways these practices can be interpreted as disruptions – in terms of how we understand both art and the city.

Note

1   Appadurai's five scapes are: ethnoscapes, mediascapes, technoscapes, finanscapes and ideoscapes and are organized based on the movement of specific things like people or ideas.

References

Appadurai, A. (1990) 'Disjuncture and difference in the global cultural economy'. *Public Culture*, 2 (2): 1–24.

Cameron, S. and J. Coaffee (2005) 'Art, gentrification and regeneration – from artist as pioneer to public arts'. *European Journal of Housing Policy* 5 (1): 39–58.

Chang, T. C. (2002) 'Renaissance revisited: Singapore as a "Global City for the Arts"'. *International Journal of Urban and Regional Research* 4 (24): 818–31.

Cole, D. B. (1987) 'Artists and urban redevelopment'. *Geographical Review* 77 (4): 391–407.

Elkins, J. (ed.) (2007) *Is Art History Global?* New York and Oxon: Routledge.

Gandy, M. (2016) 'Unintentional landscapes'. *Landscape Research*, 41 (4): 433–440.

Gregory, D. (1995) 'Imaginative geographies'. *Progress in Human Geography*, 19 (4): 447–485.

Grierson, E. and Sharp, K. (eds.) (2013) *Re-Imagining the City: Art, Globalization and Urban Spaces*. Bristol, UK: Intellect.

Harris, A. (2013) 'Financial artscapes: Damien Hirst, crisis and the City of London'. *Cities* 33: 29–35.

Hawkins, H. (2012) 'Geography and art. An expanding field: Site, the body and practice'. *Progress in Human Geography* 37 (1): 52–71.

Kong, L. (2012) 'Ambitions of a global city: Arts, culture and creative economy in "post-crisis" Singapore'. *International Journal of Cultural Policy* 18 (3): 279–94.

Kong, L., Chia-ho, C. and Tsu-Lung, C. (2015). *Arts, Culture and the Making of Global Cities: Creating New Urban Landscapes in Asia*. Cheltenham, UK: Edward Elgar Publishing.

Ley, D. (2003) 'Artists, aestheticisation and the field of gentrification'. *Urban Studies* 40 (12): 2527–44.

Lloyd, R. (2006) *Neo-Bohemia: Art and Commerce in the Postindustrial City*. New York: Routledge.

Markusen, A. (2006) 'Urban development and the politics of a creative class: Evidence from a study of artists'. *Environment and Planning A* 38: 1921–40.

Markusen, A., Wassall, G., DeNatale, D. and Cohen R. (2008) 'Defining the creative economy: Industry and occupational approaches'. *Economic Development Quarterly* 22 (1): 24–45.

Massey, D. (2006) 'Landscape as provocation'. *Journal of Material Culture* Vol. 11(1/2): 33–48.

McCann, E., Roy, A. and Ward, K. (2013) 'Assembling/worlding cities'. *Urban Geography* 34 (5): 581–589.

McFarlane, C. (2008) 'Urban shadows: Materiality, the "southern city" and urban theory'. *Geography Compass* 2 (2): 340–58.

Miles, M. (1997) *Art, Space and the City: Public Art and Urban Futures*. London and New York: Routledge.

Ong, A. (2011) In Roy, A. and Ong, A. (eds.) 'Worlding cities: Asian experiments and the art of being global'. Malden, UK: Wiley-Blackwell, 1–25.

Pinder, D. (2008) 'Urban interventions: Art, politics and pedagogy'. *International Journal of Urban and Regional Research* 32 (3): 730–36.

Ren, J. (2015) 'Gentrification in China?' In *Global Gentrifications: Uneven Development and Displacement*, edited by E. Lopez-Morales, H. B. Shin, L. Lees. University of Bristol, UK: Policy Press.

Robinson, J. (2011) 'Cities in a world of cities: The comparative gesture'. *International Journal of Urban and Regional Research* 35 (1): 1–23.

Roy, A. and Ong, A. (eds.) (2011) *Worlding Cities: Asian Experiments and the Art of Being Global*. Malden, UK: Wiley-Blackwell.

Said, E. (1993) *Culture and Imperialism*. London: Vintage Books.

Scott, E. E. and Swenson, K. J. (eds.) (2015). *Critical Landscapes: Art, Space, Politics*. Oakland, CA: University of California Press.

Sharp, J., Pollock, V., and Paddison, R. (2005) 'Just art for a just city: Public art and social inclusion in urban regeneration'. *Urban Studies* 42 (5–6): 1001–1023.

Wang, J., Oakes, T., and Yang, Y. (eds.) (2015) *Making Cultural Cities in Asia: Mobility, Assemblage, and the Politics of Aspirational Urbanism*. London and New York: Routledge.

Wojan, T., Dayton M. and McGranahan D. (2007) 'Emoting with their feet: Bohemian attraction to creative milieu'. *Journal of Economic Geography* 7: 711–36.

1 How colour replaces fear

Yazmany Arboleda and Nabila Alibhai

Introduction

In all three cities, Kabul, Johannesburg and Nairobi, divisions, insecurity, poverty, and inequality are rampant. In each case, the cities have spatial divisions rooted in complex histories and reinforced by contemporary fear. Through strategic collaboration and partnerships with experienced civil society actors, social practice art mobilized and involved thousands of community members to create a new solution to difficult but habitual conditions. Each orchestration aimed to provide the community with new tools to engage in dialogue, rekindle ownership of their environment and generate interactions that inspired love, inclusion and belonging. The art was designed to give voice and possibility to those who feel powerless.

The first, 'Monday Morning', orchestrated in the wake of a night's violence in Kabul, brought an unexpected message of peace in the form of 10,000 vivid, pink balloons carried through the city by volunteers. With 'Beware of Colour' in Johannesburg, decrepit, abandoned buildings – left to decay since the end of apartheid – became the centre of a conversation about the city's homelessness after pink paint was poured down their fronts. In divided Nairobi, 'Colour in Faith' brought Muslim and Christian houses of worship together to paint their buildings yellow in the name of love and tolerance (see Figure 1.4).

All three projects combined the motivations of the artist with best practices in civic engagement, community building and place making to amplify their impact. They were predicated on a theory of change that asserts that social practice art can complement community building in a unique and powerful way; helping communicate, relate, invent, organize and build confidence. When orchestrated in the public realm, they have the added capacity for inclusivity, immediacy and collective catharsis. The collective action required to deliver a public orchestration also becomes a form of peace and empathy building. This becomes ever more significant as social systems meant to preserve cohesion in politics, religion, trade and judicial institutions show their limitations in addressing emerging challenges. These shortcomings stem in part from slow processes and limited democratic representation beyond a relatively narrow group of connected individuals.

The first effect of social practice art is to link people together as participants of a common experience. Policy makers and urban strategists seek to create an experiential and emotional approach to the management and revitalization of the public realm; social practice artists are uniquely equipped to help turn those objectives into reality.[1] This takes place through 'triangulation in public space',[2] which William H. Whyte, urbanist and author of *The Social Life of Small Urban Spaces*, defines as a 'process by which some external stimulus provides a linkage between people and prompts strangers to talk to other strangers as if they knew each other'.[3] Whyte continues by explaining that people will be looking at each other just as much as what is being performed/activated on the street. The second effect of a social practice art is the collective reconsideration of an environment. Artists 'create an event to which people are free to participate, or not, observe, comment, interact and eventually sense collectively the experience that the artist has crafted'.[4]

There is also civic value in the public experience because the conversation it creates enhances freedom of expression and community participation. Moreover, creative agents can inspire others to artfully engage the world.[5] Doris Sommer, the director of the Cultural Agents Program at Harvard University, gives several examples of politicians who have used art to create change, the most impressive of which is Antanas Mockus, the former mayor of Bogotá, Colombia, whom she calls an 'international beacon of creative administration'. Philosopher and mathematician, the twice-elected mayor, knows and teaches the value of 'artful' responses to crime, corruption and violence.[6]

Lastly, there is argument for the 'invasion' of public space as a way to improve security. Orchestrations that garner crowds could arguably be described as an example of Jane Jacob's 'eyes on the street'.[7] The inclusive character of a collective public event may also impede criminality.[8] James Q. Wilson and George L. Kelling's broken window theory posits that urban disorder and vandalism, if allowed to fester, engender a cycle of crime and anti-social behaviour.[9] In the reverse of Wilson and Kelling's theory, we also see that urban order and vitality can lead to inclusive and proactive behaviour inspired by a sense of pride, belonging and therefore ownership.

Through a series of vignettes, this chapter will analyse the creation processes of the installations and their implications on transforming urban space, community cohesion, perceptions of security, cultivating a creative class and urban policy change. It covers

1   the ideation process;
2   the process of engagement;
3   choice of location;
4   aesthetic decisions;
5   collaborations and partnerships;
6   media engagement; and
7   tools for dialogue.

Case studies

Scene 1: 'Monday Morning in Kabul'

Arboleda had conceived of the project as a way of interrupting the daily routines of adults walking to work on Monday, the day of the week when their morale and nervous systems are at their lowest. On 25 May 2013, 10,000 bright pink balloons infiltrated the streets of Kabul, Afghanistan, seeping into neighbourhoods far from their source. Their mainspring was the Timur Shah Mausoleum, the resting place of the Durrani King who made Kabul the capital of the modern Afghan state in 1776, and which until recently had stood in a state of disrepair in the centre of the city.

Amidst the mausoleum's mulberry trees, more than 100 young Afghans laboured from sunset through the night filling the balloons with helium. Their backdrop was dusty hills, teeming with small houses. As the sun rose and the lights from these houses extinguished with the day, the mausoleum's garden swelled with electric pink, the balloons held in clusters in the fists of the volunteers, ready to burst through its narrow wrought-iron gates.

Hours before, a curl of smoke hung over the city from a large-scale attack on an international organization – a suicide car bomb followed by rocket-propelled grenades and sniper fire. The Taliban claimed responsibility for the attack, saying it had targeted a rest house used by the US Central Intelligence Agency. The sounds of the five-hour battle with security forces reverberated through the city as night fell. A policeman was killed in the attack and ten people were wounded. Emails shot back and forth between the balloon volunteers. The suggestion that the project should be aborted was countered by a commitment to proceed. As the attack continued, their plans solidified. 'We own this city', they wrote.

The next morning, 10,000 adults walking to work were met by young Afghans with clusters of pink balloons, each one handed in a declaration of promise, hope, happiness and beauty. The orchestration of this balloon invasion took ample preparation. Over five months, 130 youth organized themselves, approvals were received from the mayor and the Ministry of Defence, biodegradable balloons flown in, pedestrian movements studied, plans of action defined.

By noon that day, the balloons were out of their hands and seeping through the neighbourhoods of the city. Within days, the BBC, Al-Jazeera, ABC News, *The New York Times*, *The Guardian*, *The Independent* and numerous local TV, radio and print media picked up the balloon orchestration. The Internet bubbled with YouTube videos showing the streets of Kabul and happy faces of people walking to work.

There is one photograph of a soldier smiling awkwardly, balloon in one hand, Kalashnikov in another. For all the stories in the media about insecurity in Afghanistan, including the distrust of the Afghan Security Forces, this is a rare image of humanity. Another photograph shows a young Afghan woman, hair falling from under her pink scarf as she leans, balloon extended from her hand to a beggar in a burkha, whose entire body is covered in the custom blue cloth that makes a pyramid from her head down to the dirt road on which she sits.[10] The image is one of compassion and of a pluralistic vision of two Muslim women of the same town with wildly divergent life trajectories.

The youth orchestrators continue to creatively design their own futures. Their Facebook page shows that their numbers have quintupled. For all those killed in Afghanistan, whose numbers make fleeting appearances as statistics in the news, the spirit of happiness and possibility on this one unusual day on the streets of Kabul will hopefully endure. Days after the public art installation, the Taliban

Figure 1.1 Local Afghan artists and activists walking through the streets of Kabul gifting biodegradable bright pink balloons to adults on 25 May 2013.

issued a statement condemning the event, or as the *Huffington Post* called it 'an angry tirade against balloons' ('Taliban Publishes Angry Tirade Against Balloons After Kabul Art Project', *Huffington Post*, 2013). To some, this was a small victory. See Figure 1.1.

Scene 2: 'Beware of Colour' (Johannesburg)

In the artist's words:

> When I arrived in Johannesburg in early June 2014, my intention was simple: to work with a community of local artists and activists to create 'coloured' people, a project in which a people wearing a single saturated colour would come together and raise awareness of a prominent cultural issue.
>
> From the airport, I travelled to the city's business district. Despite having visited Johannesburg four times previously, I was taken aback by the lack of white people in the city. In re-orienting myself to the city, I noticed empty and decaying post-modernist to art deco to modernist buildings dispersed across the urban landscape. In conjunction with these decaying edifices lay homelessness rampant throughout the city. Yet, Johannesburg was a beacon of refuge for Africans from across the continent looking to build better lives for themselves. The juxtaposition of the beautiful, decaying, empty buildings and a population with no place to work or live was striking.
>
> In 1982, the broken window theory was introduced by social scientists Wilson and Kelling. They asked their audience to consider a building with a few broken windows. If the windows were not repaired, vandals would likely break a few more windows. Eventually, they may even break into the building, and – if it was unoccupied – perhaps become squatters or light fires

Figure 1.2 Water soluble pink paint cries out of one of the windows of Clegg House in Johannesburg, South Africa. Originally built in 1934, this Art Deco-Stripped Neo Classical building has been decaying since it was vacated by its white inhabitants post-apartheid in 1994.

inside. Through a series of meetings and tours, local artists provided meaningful feedback that lead us to conceptualize and create a plan forward: a project we titled 'Beware of Colour'.

The idea was simple: to pour hot pink paint from the windows and roofs of a number of decaying buildings downtown to literally highlight the social challenge and opportunity the buildings represent and to aesthetically make the buildings appear to be bleeding colour. Through word of mouth, news spread and all manner of artists got involved from painters to photographers to printmakers. Night after night over a period of eight weeks, groups met at midnight to turn the decaying buildings into crying, bleeding, leaking ones. Creative agents of all colours and creeds collaborated and with messy pink hands and curious minds discussed the challenges, the objectives and the ongoing narrative of the work. See Figure 1.2.

Scene 3: 'Colour in Faith' (Nairobi)

Conceived as a multi-city project, 'Colour in Faith' is a social movement that began in Nairobi in 2015, celebrating religious pluralism by painting houses of worship yellow. Yellow becomes the colour of light, the opposing colour of darkness, the colour that reminds us of the halos around saints and angels. The colour symbolizes and speaks to *love* as the most important notion in any religion. Colouring the buildings a single colour highlights the idea that there is more that unites than divides us as a people, African and otherwise. The orchestration brought faith outside of the walls of institutions into an experience of encounter, expression of common acceptance and tolerance, and offered the opportunity to those of many faiths to extend their reach beyond and across institutional walls. The historic culture of pluralism was reflected upon, honoured, and expressed in the public realm.

Once the concept was developed through discussions with religious leaders, public intellectuals and artists, the process of dialogue with priests, pastors, imams and Hindu leaders began. In three months, fourteen houses of worship had expressed interest in participating. At the three-month mark, the project hosted an interactive gallery event at the Circle Art Agency in Nairobi on 2 September 2015. All participating communities were invited, as well as regular gallery goers, and civil society organizations. At the launch, Arboleda said:

> Kenya has gotten headlines all over the world as a place where terror and darkness live, because of the attacks that happened in Garissa and Westgate (Mall) … and I think it's time to turn those headlines around and talk about what's actually beautiful about people who live here and actually do a lot of good for other people. And so this is about highlighting those people and those communities. The idea is very simple, it is to get people to come together, paint together, live together and to say that love is more important than anything else.
>
> (Circle Art Agency, 'Colour in Faith' Launch, September 2015)

The walls of the gallery were decked with images created by Arboleda that spoke to how people experience faith through religion in Kenya, in the present day and historically. A digital rendering depicted a security guard of these building as patron saints of safety, another showcased attachment to a religious narrative from a very young age. There was also a rendering that included a historical image from the 1977 World Conference on Religion and Peace, which was held in Nairobi. Some images from as far as the 1920s hinted at the relationship between colonialism and religion in Africa. In addition, there was a yellow mural that took over an entire wall of the gallery, and asked visitors to answer the question 'Where does your faith live?'

At around the same time, a partner organization, Fatuma's Voice hosted a poetry jam and tweet chat that engaged more than 120 poets, musicians and artists on the theme 'Colour in Faith'. This generated additional support for the project, which lead to its nomination for a Disruption by Design award given by a local cell phone company and a feature on the project in the regional paper *The East African*.

At the six-month mark, twenty houses of worship had committed and the hosting organization, in COMMONS, had raised enough funds to paint three buildings. Colour in Faith then partnered with a paint company, Sadolin Paints, for a donation of paint and the East African Institute to host five tweet chats on the relationship of faith with leadership, radicalization and patriotism. The first four buildings were painted in the summer of 2016. Muslims and Christians came together and painted while entertained by poets. Painting the buildings represented a catalytic

Figure 1.3 Muslims, Christians, and people of other faiths work together in Kibra to paint Holy Trinity Anglican Church yellow as an expression of shared humanity.

event that expressed the progress in community unity, belief in love and an expression of inclusivity as a result of the project. Local TV stations featured the project and its importance in creating cohesion and mitigating conflict in advance of the 2017 presidential election. As resources were mobilized for more buildings throughout 2016, a new partnership has been developed with Nairobi City Hall for people of all faiths to co-create a collective space in the centre of the city.

The project was also covered by the international press as a creative way to address difficult issues pertaining to religious pluralism, tolerance and our global battle with radicalization. This included features by CNN, BBC, *The Guardian*, *Huffington Post* and some significant art and religious press outlets. As a result, more communities caught wind of the movement and invited the Colour in Faith team to paint four more buildings in Likoni, a coastal town on the Kenyan coast with a history of violence. This brought the total to eight buildings by September 2016 with additional requests. See Figure 1.3.

Discussion: The crystallization of a method

Although the methodology varied, there was a basic four-step process that we developed through the planning and delivery of these projects. These steps were

Figure 1.4 One of the original 'Colour in Faith' concept drawings. 'Colour-In-Church',
graphite and water colour on paper, 2014.

intended to create empathetic collective leadership, mobilize social action and
visually imprint co-created collective aspirations. These were informed by the
leadership philosophies of Otto Scharmer[11] and the consideration of the role of
the planner as a mediator.[12] The first step was ideation, which entailed putting an
artist at the centre of building a common intent within a community. This involved
listening, observing and examining spaces. Once spaces of high potential for
evolution were identified, the next step was to create spheres of engagement or
influence, through which dialogue and personal growth takes place. This phase
takes the longest time, and involves getting buy-in from stakeholders, and creating dialogue platforms (physical, communal and virtual). The third phase involved
co-creating the art installation, which is a way to crystalize a common intention,
and create a prototype that explores the future by doing. The fourth and last phase
'solidified learning by evolving'. This evolution is the change itself – the community or networks created, the policy change, or the transformation of a workspace
or neighbourhood.

It is instructive to consider the ideation process for all three installations.
The idea for 'Monday Mornings' was originally conceived by Arboleda in 2011
in Bangalore, India, and was inspired by the notion of disrupting a morning
commute, the time when one's morale is at its lowest, by handing out 10,000
bright orange balloons to adults walking to work. The intention was for these
adults to reconnect with the local social fabric and environment. The idea was
to disrupt a mundane, habitual event with joy and celebration. The installation

was then repeated in Yamaguchi, Japan, in 2011 with green balloons. When the installation was brought to Nairobi in 2011, a grenade attack shook the centre of the city a week before the installation was due to take place. The activists were asked whether it was prudent to continue, and they felt that it had in fact become even more important. They wanted to send the defiant message that life would go on despite the atrocities, and that the city belongs to its inhabitants, not those who want to tear it apart. This inspired Arboleda to bring it to Kabul, another city under frequent assault from terrorism, and where there is even more of a need to create multiple narratives of the city beyond its association with fear and despair. A particular motivation was to honour those who live, work and create culture. In addition, there was a desire to instil hope, and to create a positive relationship between individuals and the city, both locally and in the international eye.

'Beware of Colour' was also developed through a collective consideration of social justice issues associated with buildings in Johannesburg that had been abandoned, unused and decaying for twenty years after the end of apartheid. Arboleda said:

> What's challenging in South Africa is that, although there has been so much evolution in the last twenty years, there still remain a lot of tension and divisions. This kind of project allows for space to dissect and really understand what's happening between people from the perspective of politics, skin colour, religion, and other ways in which we define identity.

The abandoned buildings in a city experiencing rampant homelessness and housing crisis were emblematic of a broken government system. The project visually highlighted these spaces as a catalyst for the community to wake up, ask questions, act up, bring these buildings back to life, and in turn make Johannesburg's Central Business District (CBD) neighbourhood a safer place for all.

'Colour in Faith' was initiated through discussions facilitated by inCOMMONS, a Kenyan civic engagement and place-making organization. Through these discussions between Kenyans, the artists unearthed growing national anxiety around terrorism, and the negative reframing of faith. Kenya in recent years has been in the global limelight as the target of fundamentalist voices and acts of terror justified on religious grounds. This is particularly sad because Kenya has a long-established culture of religious acceptance, tolerance, accommodation and exchange. This culture is undermined by an infusion of hard-lined interpretations of faith, and by the deepening of a global identity based on media stories about division, terrorist attacks and insecurity. These discussions underlined a fear that cultural polarization as a result of these attacks would have intolerant fanatics succeed in dividing societies.

The call was for an intervention that would facilitate a form of inclusive communication, and that would allow those who believe in acceptance, love and harmony to express these values, and reinforce them within themselves and with others. The call was also for collective action and strengthening of community. The result was a combination of civic engagement and peace-building with art, where a movement reinforcing religious pluralism would be created, culminating in a visual expression of love and harmony, through painting participating houses of worship yellow.

'Beware of Colour' highlights the effort that goes into identifying appropriate spaces for the installations. The process involves engaging stakeholders and a thorough analysis of pedestrian movements, visibility and impact. With 'Beware of Colour', Arboleda went through downtown Johannesburg and noticed that there were buildings, ten to twenty stories high that were boarded up and decaying. He then partnered with John Wood, a photographer who had been working on a project photographing illegal immigrants squatting in some of these buildings, and began visiting some of these buildings with him. Together they identified buildings for the installation. They considered focusing on buildings where the squatters resided, but decided that this might call attention to their illegal status, and potentially put them at risk of losing their homes and being deported. They decided to choose buildings without squatters. They then approached developers under the guise of proposing fashion photo shoots in the buildings, to get access into these buildings, photograph the spaces and assess safety and circulation patterns. Some of the buildings had been burnt through from the inside and the flooring was unstable. Many had debris from walls caving in and deteriorating materials. Nine buildings were chosen, based on access, safety, location within the neighbourhood, and how historically iconic they are.

After the idea was formed, the second phase was to collectively define the spheres of influence and establish partners. The partners influence the scope of the project and ensure longevity of impact beyond the duration of the catalytic art production phase. Partners also enable political and community access, and increase potential for expansion. Having a diversity of partners also ensures that the intended impact of the project is culturally appropriate, relevant and timely. 'Monday Morning in Kabul' was crowd-funded through a digital mural of international supporters, which connected each donor ($1 per person for one balloon) with the recipient of the balloons. More than 2,500 people from forty-two countries contributed $1 and were connected to the project in the hope of creating a better understanding of and communication with the people of Afghanistan.

For 'Monday Mornings in Kabul', we approached civil society in general and artist collectives in particular that became involved in the process, providing human resources and an artistic context to explore together (see Figure 1.5). Approvals were sought from the mayor and the Ministry of Defence, biodegradable balloons were flown in, pedestrian movements studied, plans of action defined. The 130 youth volunteers convened at the Timur Shah Mausoleum, access to which was granted by the Aga Khan Trust for Culture, the organization that had restored that national heritage site.

The wide array of local partners participating in 'Monday Mornings' included: the Centre for Contemporary Arts Afghanistan, The Faculty of Fine Art of Kabul University, Afghan Women's Network, Young Women for Change, AEISEC – Youth Leadership, AIS: Afghan International Services, Sale Foundation, and art collectives Kabul Street, Berang, Roya Film House and Parwaz Puppet Theatre. These organizations' members helped organize and execute the

Figure 1.5 One of the original 'Monday Mornings in Kabul' concept drawings. 'Untitled'
353 x 500 mm, graphite on paper, 2012.

installations, and made it possible to get the necessary approvals from government authorities. Moby Media Group and Film Annex also agreed to sponsor the project, providing financing and other resources. International support came from the United Nations Assistance Mission for Afghanistan (UNAMA),[13] and the Dutch Embassy.

The Facebook page created for the project united all participants with supporters around the world.[14] It continues to function as a hub for information where participants share available jobs, new civic initiatives, and other relevant data that could benefit other members of the community. Once the installation had taken place, dialogue was amplified by coverage in newspapers, and websites all over the world. *The New York Times* covered the installation on the front page of its website, as did ABC News,[15] news.com.au,[16] Berlin's *Battery Magazine*, the *Huffington Post*, *Domus* magazine, IPI Global Observatory, *The Jakarta Globe,*[17] *Agence France Presse*, and other news outlets numbering about twenty local and international articles. It was also covered through video by Tolo TV and even the BBC.

Similarly, the other installations have engaged local artists and activists and local civil society organizations, academic partners and social media. 'Colour in Faith' has used Instagram, Facebook, WhatsApp groups and Twitter, the latter of which has been used to combine academic research and dialogue to create conversations around the relationship between faith, leadership, patriotism, and radicalization.

The third phase, execution of the projects, had the effect of disrupting habitual patterns of existence having participants reflect on current social conditions with

Figure 1.6 Thought of as a 'Living Sculpture' by the artist, Jeddah Mosque Kambi in Kibra (the largest informal settlement in Africa) is one of the buildings painted for 'Colour in Faith' in 2016.

Figure 1.7 One of the original 'Monday Mornings in Kabul' concept drawings. 'Untitled', 353 x 500 mm, graphite on paper, 2012.

the intention of envisioning a better social reality. This increased awareness of long-standing and often dysfunctional perceptions and ways of relating to others. 'Colour in Faith' calls on individuals to reconsider their association with faith-based institutions, so that their affiliations are based on furthering common humanistic values rather than exclusive membership in predetermined groups. Globally, religions are undergoing transitions. Their relevance to youth and to contemporary life was questioned, and revealed that religious leadership struggles to continue to appeal to the young in particular.

These installations promoted collaboration between groups that generally did not work with one another, and increased awareness of individual and collective identities, and their socio-cultural influences. Like any community event, these installations generated more than co-operative action, they also created a network of relationships that ripple into other projects.[18] 'Monday Mornings in Kabul' established a community of 130 youth volunteers who, after collaborating on the installation, went on to collectively volunteer in soup kitchens, to create a series of documentaries on artists in Afghanistan, and to share job postings on their Facebook page. 'Colour in Faith' established new relationships between Christian and Muslim communities in the neighbourhoods of focus. In addition, the local administration, the Office of the Minister of Parliament, used 'Colour in Faith' to promote additional inter-faith activities. 'Colour in Faith' also engaged an academic partner, the East Africa Institute, to enhance the visibility of the project and to disseminate their research on identity, civic agency, leadership and corruption and engage Kenya and the content in related tweet chats.

Moreover, they created an alternative international narrative about these cities. Each one highlights the people in these cities who create culture and beauty, rather than the terror that they are known for. In the case of 'Monday Mornings', Arboleda says: "I see this project as a platform that transforms the single story of catastrophe that the world sees in Afghanistan into multiple narratives that highlight our shared humanity."[19]

Internationally, the press showed images of Kabul that were diametrically different from images conventionally depicting fear, insecurity and war. Instead, a powerful image of the people of Afghanistan was created with a more normalized and balanced understanding of the presence of all types of people like pharmacists, fruit-sellers, taxi drivers, musicians and artists.

The processes also introduce individuals to alternative and improved ways of communicating positive aspirations. This was done through the process of art making and engaging with others in unusual ways – both physically and through social media. More broadly, the co-creative art process allowed socially conscious actors to reach a wide audience with messages of positive change, social integration and peaceful co-existence. Strengthening communication helps mitigates anger and violence. This is achieved through providing people with non-violent means to communicate differences and commonalities. The interventions have in common that they engaged people in exercises that are creative, hands-on, energizing, stress-reducing and morale building.

This process of co-creation has a potential impact on extremism, since it strengthens positive networks between youth, and between youth and their elders. According to anthropologist Scott Atran 'about 3 out of every 4 people who join Al Qaeda or ISIS do so through friends, most of the rest through family or fellow travellers in search of a meaningful path in life'.[20] He points to the reality that parents are rarely aware of the decisions of their children to join radical movements. This is indicative of a world order where horizontal peer attachments are more powerful than intergenerational vertical lines of attachment. In his address to the United Nations Security Council in 2015, he goes on to outline three things that young people need: something that makes them dream of a life of significance through struggle and sacrifice in comradeship; a positive personal dream, with a concrete chance of realization; the chance to create their own local initiatives. All three projects 'Colour in Faith', 'Beware of Colour' and 'Monday Mornings' fulfil these conditions.

The projects also managed to lift the profile of art's ability to inspire social change and leadership. In Kenya, 'Colour in Faith' collaborated with a poetry collective, Fatuma's Voice. The launch of the project took place at the Circle Art Agency, which brought artists, art enthusiasts and religious leaders into the same space. At the opening, the artists designed an interactive wall that asked the question 'Where does your faith live?' that got people to reflect on their values.

A poetry jam brought together a community of 120 artists and activists to feed into a national conversation about how we think about faith in our collective space. Over twenty participants volunteered to help drive the project forward through community organizing, local fundraising, engaging houses of worship, and publicizing the project through art and painting. A local events magazine hosted a panel discussion with 'Colour in Faith' on 'Multi-culturalism: Can Art Shift Culture?', which was preceded by an article in their paper on the same topic.[21]

The last phase is the reinforcement of the result. It is the crystallization of the evolution of a community through the art-making process. This impact can extend beyond the inter-personal transformations and the resulting dialogue described above. Changes are seen in the physical environment, perceptions of security and belonging, and in some cases policy changes. With 'Beware of Colour', community members verbalized how 'broken windows' lead to other types of vandalism and disrespect for public spaces around the neighbourhood. The fluorescent intervention shifted how passers-by, neighbours and the government thought about their environs and their potential. In fact, the dialogue that ensued after the installation resulted in the city moving forward with a plan to renovate the entire neighbourhood.

In Afghanistan, because of the suicide and gun attack that had taken place the night before the installation, the project took on a new meaning. As Arboleda explained: 'I did think of calling it off that night, but all the volunteers insisted it continue. I could hear explosions from my house, but everyone was just ignoring them and doing the last-minute preparations undeterred'. The volunteers saw the project as a means of reclaiming the city for themselves and having it be a statement about whom Kabul really belongs to – not to the Taliban but to them. Also, the women that participated were emboldened, and in turn emboldened other women to walk and be seen in the city.

Most importantly perhaps is the impact of the installations on hope and agency. The process of art making in this way is a demonstration of creative problem solving. It is the use of the imagination to envision a better future and it is an exercise in organization, planning and execution. Being able to take communities from a process of self-reflection, to connection, to collective prototyping to action can have an incredible impact on their ability to accomplish almost anything.

Conclusion

The artistic interventions have been able to accomplish many of the aspirations of peace-building activities. This includes changing perceptions, strengthening communication, addressing anger and violence and increasing cooperative action. The basic philosophy behind the art practices described has been to integrate art-making and civic engagement to create an experience of collective problem solving, catharsis, relationship-building, and, through their visual impact, to be able to reach wide audiences with messages of value-driven living, social integration and healthy coexistence.

The aspiration, to borrow from Scott Atran, is to create

> a global archipelago of such peace builders ... what is most important is quality time and sustained follow-up of young people with young people ... involving not just entrepreneurial ideas, but also physical activity, music and entertainment ... It takes a dynamic movement that is at once intimately personal and global.[22]

These installations, in addition create a visual living memory of positive collective aspiration.

Furthermore, these engagements support artists, elevating their importance as culture-builders and change-makers. The idea is that the imagination, and imagining with others, is really the necessary component of changing and innovating in any way. Arboleda refers to art not as a noun but as a verb. Being an artist is having the capacity to engage, process, reinvent and reframe. In redefining art in this way, it makes it possible for everyone to be an artist. This capacity is relevant to every aspect of our lives. And the process of art making, that is to reconsider our place in the world, to improvise, to entice others into a creative experience of our own making – is to participate in the design of how we relate.

The collaborators have created two start-ups that espouse the methods described for implementing these installations. The first is limeSHIFT, a US-based company that uses art to make communities – private and public – their most creative, productive and connected. Essentially taking community building and peace-building practices and, through art, engaging private sector companies to steer them through a process of collective reflection on their values, then assisting them to visually manifest these values within their working environments and in their neighbourhoods. The second is inCOMMONS, a Kenyan-based company with the mission of engendering tangible and personal responsibility for public spaces, culture and the environment. inCOMMONS focuses on creating spaces that nurture shared responsibility for our collective bounty. Both companies exchange artists and methodologies, creating an innovative north-south collaboration.

The common approach is to create empathy that translates into purpose followed by action. The whole process is facilitated through art-making such that a collective intention is visually manifested through beauty and participation. The impact is the activation of the imagination of many, and a call on us to collectively and deliberately create an expression of a world order that is better than the existing one.

Notes

1 Vivian Doumpa, Nick Broad (2014). 'Buskers as an Ingredient of Successful Urban Places', *Future of Places*, Buenos Aires.
2 As introduced by William H. Whyte and Project for Public Spaces.
3 Whyte, W. H. (1980). *The Social Life of Small Urban Spaces*. Washington, DC: The Conservation Foundation.
4 Whyte, W. H. (1980). *The Social Life of Small Urban Spaces*. Washington, DC: The Conservation Foundation.
5 Sommer, D. (2005). *Art and Account, Review: Literature and Arts of the Americas*, Issue 71, Vol. 38, No. 2, 261–276.
6 Ibid.
7 Jacobs, J. (1961). *The Death and Life of Great American Cities*. New York: The Modern Library.
8 Landry C., Greene L., Matarasso F. and Bianchini F. (1996). *The Art of Regeneration: Urban Renewal through Cultural Activity*. Glos: Comedia, The Round, Bournes Green, Stroud.
9 Wilson, James Q., Kelling, George L. (March 1982). 'Broken Windows: The police and neighborhood safety', *The Atlantic*.
10 www.myvoicetv.com/2013/06/a-volunteer-gives-pink-balloon-to-woman.html.
11 Otto Scharmer (2009). *Theory U: Leading from the Future as it Emerges*, Oakland, CA: Berret-Koehler Publishers.
12 'Mediated Negotiation in the Public Sector: The Planner as Mediator', *Journal of Planning Education and Research*, August 1984 (4): 5–15.
13 unama.unmissions.org/10000-pink-balloons-dot-afghan-capital-'oneness-and-peace'-spring.
14 www.facebook.com/WeBelieveInBalloons.
15 abcnews.go.com/International/artist-hand-10000-pink-balloons-kabul-afghanistan/story?id=18648284.
16 www.news.com.au/world/after-attack-yazmany-arboledas-peace-balloons-bring-hope-to-afghanistan/story-fndir2ev-1226650628723.
17 jakartaglobe.beritasatu.com/features/after-kabul-attacks-10000-peace-balloons.
18 Page156, 'The Great Neighborhood Book', Project for Public Spaces.
19 www.huffingtonpost.com/2013/05/25/we-believe-in-balloons-day-artist-yazmany-arboleda-kabul-afghanistan_n_3335878.html.
20 Scott Atran, address in the UN Security Council's Ministerial Debate on 'The Role of Youth in Countering Violent Extremism and Promoting Peace', 23 April 2015.
21 www.upnairobi.com/2016/06/05/q-a-with-nabila-alibhai-colour-in-faith-multiculturalism.
22 Scott Atran, address in the UN Security Council's Ministerial Debate on 'The Role of Youth in Countering Violent Extremism and Promoting Peace', 23 April 2015.

References

AFP (2013, May 25) 'Day after Kabul attacks, 10,000 peace balloons'. Retrieved from jakartaglobe.beritasatu.com/features/after-kabul-attacks-10000-peace-balloons.

Curry, C. (2013, March 11) 'Artist to hand out 10,000 balloons in Kabul, Afghanistan'. Retrieved from abcnews.go.com/International/artist-hand-10000-pink-balloons-kabul-afghanistan/story?id=18648284.

Jacobs, J. (1961) *The Death and Life of Great American Cities*. New York: The Modern Library.

Landry C., Greene L., Matarasso F., Bianchini F. (1996) *The Art of Regeneration: Urban Renewal through Cultural Activity*. Glos: Comedia, The Round, Bournes Green, Stroud.

News.com.au (2013, May 26) 'After attack Yazmany Arboleda's peace balloons bring hope to Afghanistan'. Retrieved from www.news.com.au/world/after-attack-yazmany-arboledas-peace-balloons-bring-hope-to-afghanistan/story-fndir2ev-1226650628723.

Scharmer O. (2009) *Theory U: Leading from the Future as it Emerges*. San Francisco: Berrett-Koehler Publishers, Inc.

Skibola, N. (2013, May 21) 'We believe in Kabul'. Retrieved from www.huffingtonpost.com/nicole-skibola/we-believe-in-kabul_b_2904136.html.

Sommer D. (2005) 'Art and Account', *Review: Literature and Arts of the Americas*, Issue 71, Vol. 38, No. 2, 261–27.

Susskind L. and Ozawa C. (August 1984) 'Mediated negotiation in the public sector: The planner as mediator'. *Journal of Planning Education and Research*, 4: 5–15.

UN Security Council's Ministerial Debate on the Role of Youth in Countering Violent Extremism and Promoting Peace (2015, April 23) Address by Scott Atran. Retrieved from: blogs.plos.org/neuroanthropology/2015/04/25/scott-atran-on-youth-violent-extremism-and-promoting-peace.

UNAMA, Kabul (2013, February 18) '10,000 pink balloons to dot the Afghan capital for "Oneness and Peace" this spring'. Retrieved from unama.unmissions.org/10000-pink-balloons-dot-afghan-capital-%E2%80%98oneness-and-peace%E2%80%99-spring.

UP Magazine (2016, June 5) 'Colour in Faith', Q & A with Nabila Alibhai. Retrieved from: http://www.upnairobi.com/2016/06/05/q-a-with-nabila-alibhai-colour-in-faith-multiculturalism.

Vivian Doumpa and Nick Broad (2014) 'Buskers as an ingredient of successful urban places, future of places', Buenos Aires.

Whyte, W. H. (1980) *The Social Life of Small Urban Spaces*. Washington, DC: The Conservation Foundation.

Wilson, James Q. and Kelling, George L. (March 1982) 'Broken Windows: The police and neighborhood safety', *The Atlantic*.

2 The collective moment

Post-Museum's Rowell Road period

Jennifer Teo and Woon Tien Wei

Introduction

In 2008, Singapore Art Museum presented *Artists Village: 20 Years On*, an exhibition that aimed to show The Artists Village's (TAV) work in its first twenty years. Co-author Woon Tien Wei was involved in the exhibition as a member of the curatorial team. TAV is a contemporary art group the goals of which are to promote contemporary art and to bring about a better understanding of contemporary art practices and their contribution to society (*The Artists Village: About Us* 2005). The group emerged in 1988 with the establishment of an artist colony in a chicken farm in a Kampung[1] in Ulu Sembawang. They lost the 'village' space in March 1990 when the land was repossessed by the Singapore government for urban development. Despite losing the physical space, TAV continued.

TAV has exhibited in museums and art institutions in Singapore and internationally. Many curators and writers have acknowledged the importance of TAV in relation to the development of contemporary art in Singapore and that its emergence in the 1980s marked the beginning of contemporary art and would change the way art is made (K. C. Kwok, 1996; Storer, 2007; E. Tan, 2007; Turner and Barclay, 2005). Yet there has never been an attempt to survey the work of TAV in an exhibition. *The Artists Village: 20 Years On* was the first attempt to showcase the group through a retrospective and to locate its position in art history. However, the exhibition was a modest attempt at that. Sociologist Kwok in his introductory essay to the catalogue said: 'The exhibition does not attempt to wrap things up and tie loose ends; instead it unwrapped two decades of development of the group' (K. W. Kwok, 2009: 1–3).

While this was a significant attempt to unpack two decades of the arts group, the exhibition recalls one of the challenges of curating the story of collectivity in art. The exhibition was staged in two of the museum's lower galleries, which was too small to sufficiently show the twenty years of the group's history. The exhibition showed artworks which were made between 1988–2006 by artists who were affiliated with the group, with most of the artworks shown made by individual artists except for three pieces, which showed collaborative projects by the artists in the post-Ulu period (2000–2008). These pieces were *The Bali Project* (2001), *B.E.A.U.T.Y.* (2002) and the *Public Art Library* (2003). The physical artworks were centre stage, while the interactions between its members, friendships, art strategies, documentation work and other cultural activism-type activities of TAV took the form of captions, time-lines, charts and documentation that served as background information. This format suggests that these activities are peripheral and serve to only support the artworks within the exhibition, whereas we view that these are essential in understanding TAV. They make a critical contribution to TAV's identity and contribution to contemporary art in Singapore.

As curator and art critic Okwui Enwezor pointed out, collectivity and collaborative practices generate critique and question the modernist reification of the artist as an autonomous individual within modernist art (Enwezor, 2006). He raised three issues which problematise collectivity within modernism. The first is the issue of the authenticity of a work of art, as collective work complicates modernism's idealisation of the artwork as the unique object of individual creativity. Second, as collectivity is often a response to crisis, the nature of collectivity often extends into the political horizon. This tends to give collective work a social rather than artistic quality. Hence, collectivity is often seen to be 'essentially political in orientation with minimal artistic instrumentality', challenging modernist formalism's insistence on the primacy of the artwork. Third, collectivity can also be understood as a critique of the reification of art and the commodification of the artist. Under the operative conditions of capitalism, the loss of the individual artist is undesired, thus collectivity inherently rejects capitalism, and capitalism rejects collectivity.

In that sense, locating and positioning the work of art collectives and collaborative practices in art history is never straightforward and easy. The modernist art framework is deeply invested in the figure of the genius artist as the dominant thread in the art historical narrative that subsumes everything else. Hence, the experience encountered in *The Artists Village: 20 Years On* is not unique to Singapore but is a limit of the modernist art framework. To understand the nuanced complexity in the operation of art collectives and appreciate their contributions, it is important to broaden the framework of looking and reading the work of art collectives and art groups.

'Post-Museum' is an independent cultural and social space in Singapore, which aims to encourage and support a thinking and pro-active community. It is an open platform for examining contemporary life, promoting the arts and connecting people. Through their social practice art projects, Post-Museum aims to respond to its location and communities as well as find ways to create micro-utopias where the people actively imagine and create the cultures and worlds they desire. Currently operating nomadically, Post-Museum continues to organise, curate, research and collaborate with a network of social actors and cultural workers.

This chapter examines the Rowell Road period of Post-Museum (2007–2011), which was instrumental to Post-Museum's finding its first 'frame'[2] and gave shape to their social practice. This period developed and provided a co-constitutional space for the arts and civil society where artists, activists and civic actors met, a space they could push boundaries and be free to engage with social and political issues.

Figure 2.1 Post Museum on Rowell Road (Photo courtesy of Post-Museum).

Building Post-Museum: Rowell Road period (2007–2011)

In 2004, the authors co-founded 'p-10',[3] an independent curatorial team based in Singapore, to address the uneven ground for art. The art environment at that time seemed to over-emphasize the production of artworks and exhibitions, while putting little focus on the development and discursive aspects of art.

Because of the work of p-10, a businessman approached us with an offer to rent his Rowell Road shop-houses (107 and 109 Rowell Road) in Little India, an exciting historical and multi-cultural area in Singapore.[4] At that time, the properties were being rented out to a brothel and a Karung Guni shop.[5] The shop-houses and the rental were about four times that of the space that p-10 was renting, hence undertaking the new space would require substantially more resources, a common challenge for many independent art spaces. It is important to highlight that the businessman's generosity in reducing the rent and bearing a portion of the renovation costs convinced p-10 to undertake the challenge of operating a new arts space in the shop-houses,[6] and effectively made him its patron. After discussions within p-10, it was decided that the new space would be run by the authors of this chapter.

From 2000, Singapore underwent a period of rapid cultural development and liberalisation introduced through new cultural and social policies by the Singapore government.[7] The 'Renaissance City Master Plan' most clearly articulates the government's commitment to this change and how it was motivated by Singapore's move towards a knowledge-based economy (Ministry of Information Communications and the Arts, 2008; Ministry of Information and the Arts, 2000). Through the Renaissance City Plan (RCP), the government has become more involved in cultural development, through increased arts funding and the building of infrastructure and institutions. At the same time, the cultural and social policies introduced from 2000 have led to Singapore opening up and becoming more liberal as the government loosened its laws and allowed more freedom of expression. The Singapore government implemented a series of liberalisations in the form of casinos, bar-top dancing and later opening hours of clubs and pubs (Lim, 2009). They liberalized political spaces through the creation of Speaker's Corner.[8]

However, these steps seem merely 'gestural' as the government remains cautious of 'disruptive' works, and continues to exercise control over the outcome of artworks and programmes through various methods, including registration, funding, licensing and censorship (Lee, 2005, 2007; Ooi, 2010). In addition, the RCP, in envisioning Singapore as the vibrant global city of the arts, seems to define and limit art's role as market-driven, decorative and non-disruptive (Chua, 2008; Kong, 2000).

Post-Museum was a response to the RCP, our experiences working with p-10 and a general sense of frustration with the crisis of a developing Singapore arts scene. First, government-led cultural development[9] efforts led to an increase of exhibition spaces and an emphasis on exhibitions and mega-art events. Second, government-led cultural development increased the visibility of the arts but did not manage to raise art's position in society. Third, community-led cultural development[10] seemed to be decreasing as there were fewer independent initiatives from the community.

Considerations in setting up the space

Post-Museum's direction, business model and the design of the space were finalised after consultations and discussions with the patron, architect and other artists.

Two major considerations for the direction of Post-Museum were that it should be independent of government funding and that it should be a space that was conducive to activities that address the crisis of the Singapore art scene. Receiving public funds for the space and activities would require Post-Museum to produce a number of projects while limiting the scope of its activities, and we wanted the autonomy and flexibility of deciding what to focus on and how to work. In fact, we were particularly interested in the potential of non-market-driven, non-decorative and disruptive practices to expand the role of art in Singapore.

There were two important strategic decisions in terms of the business model and space design: registering as a commercial entity and changing the use of the premises. Instead of registering as an arts society or a non-profit organisation, which was often the case with arts initiatives in Singapore, Post-Museum registered as a private limited company. This reflected its goal of being a self-sustaining social enterprise, with its main sources of income coming from the sale of artworks, management of events and rental of spaces. In addition, as a commercial entity, it would have more freedom than a society or non-profit organisation in how it conducts its business. The premises were designated for commercial use under the Urban Redevelopment Authority land use plan, which meant it could be open to the public all the time (Urban Redevelopment Authority, 2003), unlike the previous p-10 space, which had limited access as it was designated as a residential space. We also applied for a legal change of use (to a restaurant) for one of the shop-houses as the food and beverage space was to be an important component in Post-Museum.

Post-Museum was officially opened in September 2007 as an independent cultural and social space, which aimed to encourage and support a thinking and pro-active community. It functioned as an open platform for examining contemporary life, promoting the arts and connecting people. The premises included the Show Room (multi-purpose space) and Food #03[11] (deli-bar and artwork of author Woon Tien Wei) on the ground floor, which were accessible to the public. The second floor and third floors were semi-private spaces, which consisted of a shared office space, artist studios and the Back Room (multi-purpose space). With the broad mission statement and various spaces available for use, Post-Museum set up an open system where almost anything was possible.

Post-Museum was located on Rowell Road in an area that was one of the oldest surviving red light districts in Singapore (Ho, 2013). Besides the illegal trades,[12] the area also had legitimate businesses and residential spaces. The landlord evicted an operating brothel and Karung Guni shop so that the Post-Museum could set up there. We felt that it was important for Post-Museum and its audiences to acknowledge the history of the building. We wanted the programmes of Post-Museum to respond to and engage the existing communities.

This could be seen in the first three projects of Post-Museum. The first was *Building Post-Museum* by Chua Koon Beng. Chua was commissioned to draw portraits of the people who were involved in the actual construction of Post-Museum. These included the migrant workers who built the space, the landlord, the staff of Post-Museum and the authors. Chua's portraits were captioned with their names as a form of recognition to the people who made Post-Museum possible. Everyone, including the migrant workers, were invited to the exhibition and author Jennifer Teo mentioned that 'their smiles really made' her day (Benjamin, 2008). The second project, *Eternity* by Chua Chye Teck, showed a series of photographs, which Chua shot of the emptied brothel after it vacated the premises, before the construction of the new Post-Museum began. The third project was a presentation by Dennis Tan, entitled *Designing Post-Museum*, which covered various design considerations involved in transforming a 1920s shop-house into a cultural space. These first three projects, which took the forms of the processes, exhibitions and dialogue, reflected Post-Museum's intention to engage its immediate public and larger community through an intentional process of collectively remembering the past, looking at what's happening now and imagining the future.

In the design of Post-Museum, the deli-bar, Food #03, was envisioned as an important component of Post-Museum. Food #03 offered an affordable menu and a space for the community to hang out. It ran a series of programmes and contained a community board where people could put up promotional posters of their events as well as a shop selling merchandise from artists and civil society groups. Food #03 provided a conducive environment with the potential for the community to engage in social projects, push boundaries, experiment and be free to engage with social and political issues, which state-led institutions would often discourage. One of the major projects started in Post-Museum took place in Food #03 on Mondays when it was officially closed: The Soup Kitchen Project (Singapore). Volunteers would cook in Food #03's kitchen, pack the cooked food then distribute it to the needy in the area.[13] Even though Food #03 has closed, the Soup Kitchen Project still exists today (albeit in a different form).

Art within cultural work

We have always held the belief that the arts and culture are important and relevant to life. However, despite the increased visibility of art in Singapore's 'renaissance', art's position in Singapore remains relatively low. Although artists and art groups continued to claim that their art was engaging with the public and society, there was little evidence to support that. As mentioned, the p-10 model of operation offered limited access to the public due to the residential status of their premises. Previous experiences seemed to show that independent art events and exhibitions reach out to a small community.

The concerns raised here are not new. There have been similar ambitions and concerns that resulted in community-led initiatives. This is best reflected in a statement by theatre doyen and founder of The Substation,[14] Kuo Pao Kun, on why he wanted it to be a 'home for the arts':

> We don't have an arts centre in Singapore. We have theatres and art galleries but these are all places where you cannot really 'stay'. There was no way you could mix with the artists. [...] I am very concerned about creating a space in Singapore life for the arts. A space not in terms of a place, but a space in our value systems, lifestyle and consciousness. A space that will be as important in our lives as the need to find a job [...] The Substation will be a permanent space to do arts, see arts, talk arts and live the arts. A space where artists can mingle and encounter strange artistic activities far removed from their own.
>
> (Sasitharan, 1990)

As illustrated through Kuo's statement, Post-Museum's emergence is not a response to an entirely new concern but to a persistent concern that continues in Singapore's 'renaissance'.

In this way, the forming of Post-Museum has a cultural-activist motivation of offering public greater access to arts and culture. One of the most prominent strategies adopted was to position art within the broader framework of cultural work. This expanded the scope which we normally worked with because we felt that as art in the last twenty years had failed to overcome the crisis, we needed to find new ways of responding. This took the form of expansions into three different areas. First, the inclusion of other forms of the arts and creative work, like music, dance, theatre, film, architecture and design. Second, the inclusion of thinkers, teachers, writers and researchers, both academic and independent, was critical in the interest of education and the development of research and critical thinking. Third, the inclusion of civil society work was part of our definition of cultural work.

The Singapore government's vision of 'active citizenship' encouraged Singaporeans to play pro-active roles in building society through promoting the notion of civil society in the early 1990s (Koh and Ooi, 2004). The process of liberalisation and opening up of Singapore has seen an increased presence of civil society groups. However, the liberalisation of civil society has been criticised as 'gestural' as the civic space continues to be highly depoliticised. This is because there remains the lingering presence of the ambiguous out-of-bounds markers, which continue to regulate most aspects of public life (Lee, 2005).

The inclusion of civil society in Post-Museum's cultural space would potentially allude to an expanded discursive space, which is necessary for political discourse to mature and for trust to develop within civil society and with the state. Hence, allowing 'the emergence of a beneficent civil society acting alongside the state towards some sense of common good and social cohesion' (Kuo, 2000). We believe that active participation in causes should be an important part of culture in Singapore's 'renaissance'. This definition shifts the parameters of Post-Museum's work and opens its space to a broader spectrum of users, audiences, participants and collaborators.

In addition to art-related activities, other types of activities took place in the different spaces in Post-Museum throughout this time. Exhibitions like *Go Veg* by Vegetarian Society Singapore (2007), *Integrate* by Migrant Voices[15] (2008), *Shopper/Hero*[16] (2008), *All You Need Is Love*[17] (2009) and *Activist Care Center*,[18] reflected the inclusion of civil society and their work in the oeuvre of Post-Museum from this early period. The *DocuLovers* series, which invited anyone to share a documentary they liked, and The Rowell Road Reading Group, which held regular group discussions based on Singapore-centric academic readings, are examples of activities focused on education and encouraged the development of research and critical thinking. From an overview of these activities, we can see that Post-Museum accommodated activities of broad interests and disciplines. This also reflected the co-founders' idea of an expanded/expansive cultural work.

Although founded by p-10, Post-Museum works differently from p-10, which functioned as a curatorial collective with a fixed team working in the field of fine arts. Post-Museum would be what curator Okwui Enwezor defined as a 'networked collective', where collaborators and collaborations are on a project basis rather than a permanent alliance (Enwezor, 2006). First, being founders of Post-Museum, we saw ourselves as 'custodians' of the space and adopted an open philosophy to collaborating, planning and programming. This allowed for collaborators from a diverse background, and although affiliated with Post-Museum, collaborations are on a project basis rather than a permanent alliance. Second, as a cultural and social space with an expanded definition of cultural work, Post-Museum allowed for a more diverse participation from cultural workers than limited to those from the field of fine art. This diversity may lead to difficulty in categorising or branding Post-Museum but this was intended by the founders. It needs to be stressed that art spaces have a limited scope under government-led directive. Post-Museum's status as an independent space (not receiving any government funds) and private commercial entity allowed it the flexibility and autonomy to define its activities.

Post-Museum attempted to bring together diverse groups in the hope of seeing these disparate practices find a connection with each other. As such, Post-Museum consciously planned and defined its community and encouraged people in this community to develop projects and programmes in their space. Hence, using it as a place to allow artists, film-makers, designers, researchers and activists to share their practices with the public and the community. As 'custodians' of Post-Museum, we respected the autonomy of the organisers and they were free to do what they wanted. In this way, they took their own responsibility for the success or failure of their projects. This meant that there was very little 'curation' for the organisers. We maintained this position because we felt that if the way we had been working during the p-10 period did not bring about much transformation to the art scene, we should try to start afresh and allow new possibilities.

Windows of opportunity in the 'Renaissance City'

The authors feel that the process of liberalisation in Singapore has been slow as the government has often introduced policies that seem to open up only to then make a U-turn soon after.

Singapore political theorist Kenneth Paul Tan observed that there are three positions in responding to Singapore's 'renaissance' (K. P. Tan, 2007, pp. 253–268). One is the optimist who believes that liberalisation of Singapore will happen in time as it is inevitable because the global market will force Singapore to open up. The second is the pessimist who considers that liberalisation efforts by the Singapore government are not 'genuine' because it is fundamentally an economic policy that will maintain the status quo of politics, thus retaining excessive control on its people. The third is the strategist, who desires genuine and comprehensive liberalisation, and having seen enough policy U-turns, is engaging with Singapore's 'renaissance' by working out strategies and tactics for dealing with what are perceived as windows of opportunity created by shifts in the government rhetoric, in the hope of realising a more open and liberal Singapore.

Post-Museum adopts the strategist position in engaging with Singapore's 'renaissance'. We believe that the RCP has created many opportunities and new possibilities for the arts, and that a more open and liberal Singapore is possible.

In her study of community-led independent art spaces in Singapore, Yvonne Tham observed that Post-Museum and other independent art spaces play a vital intermediary role for the development of artists and experimental artworks (Tham, 2009). While government-led development, which focused on large-scale art events, art market and museums, was limited by its position. Tham argued that independent art spaces were unique for their flexibility in mapping out a self-defined position, and these spaces initiated by artists, collectives or art administrators, offered support for artistic production, experimentation and discourse independent from the economic motivations of galleries and art fairs.

Independent art spaces often respond to the cultural conditions of the city, offering a platform for art to connect and engage with broader issues of politics, identity, culture and society. The status of independence also allows the independent art spaces the freedom to reject or appropriate market exchanges and functions, which is an option that institutions like museums and biennales cannot exercise. As such, independent art spaces are potentially the most flexible spaces and can address the needs of a contemporary artist in connecting to his/her audience, and importantly, are still able to prioritise art's cultural value over economic significance.

Although Singapore seems more open to the creativity of artists, artistic expression pushing the social and political limits remains limited. The Singapore government remains wary of the arts and regards such expressions as unproductive and discourages such practices through its funding policies. As early as 2007, several of the events which took place in Post-Museum were held there because the event organisers faced rejection by other venues due to their event being perceived as 'too political'.

Through the years, Post-Museum became a popular choice as it has gained a reputation for being open to such activities. The Show Room, which was used for art exhibitions, theatre and music performances, was also used to hold a press conference for a new political party[19] and a public forum on the general election.[20] Discussion of politics is often considered highly sensitive in Singapore and, in fact, is one of the no-go areas to arts groups. In addition, there is a general fear of being on the wrong side, and such activities often suffer from self-censorship. Hosting these events is not an attempt to politicise the space but to show that sensitive topics like politics not only has a space in Post-Museum, but should also have a space in society. Therefore, Post-Museum's gesture in allowing political activities to take place in its space could also be a re-purposing and re-imagination of the use of spaces for arts and culture. As Tham observed, Post-Museum's independence allowed flexibility and freedom to define its position. With this flexibility, Post-Museum managed to build a broad-based community consisting of cultural practitioners, educators, activists and a growing audience. This position was possible because of the 'networked collective' nature of Post-Museum and its model, which opened its premises to different interest groups including those involved in work which would generally be considered sensitive.

In another research on the cultural impact of independent arts spaces, Heather Chi observed that Post-Museum's strategy of actively fostering the arts and civic culture is crucial to 'ordinarizing' Singapore's Renaissance City vision (Chi, 2011). Singapore's Renaissance City vision is often read under the dominant narrative of 'cultural capital' (Zukin, 1995) or 'global city' (Hamnett, 1995; Sassen, 1991). Chi's thesis proposed to read Singapore as a creative city from a postcolonial perspective of an 'ordinary' city where collective actors constituted by 'distinctive assemblages of many different kind of activities' (Robinson, 2006, p. 170) have the capability to shape their own futures, even if they exist in a world of (power-laden) connections and circulations. This framework provides an understanding of how community-led initiatives and their emancipatory projects are culturally significant in providing an alternative 'Renaissance City' vision.

Hence, Chi proposed to view the Renaissance City not merely as an urban image of economic revitalization with a Singaporean flavour but rather as the 'ordinary' 'Renaissance City' vision, which reflects an urban reality always-already acculturated in multiple diverse and potentially emancipatory ways.

In that sense, Post-Museum's co-constitution of the arts and civil society offers a 'good' space to develop and construct an alternative 'Renaissance City' vision.[21] Chi observed that the conviviality and openness facilitated by Post-Museum have over time positioned Post-Museum as a visible alternative space/community, where its community felt a relative ease to organise events at Post-Museum, where like-minded people could meet and civic actors and artists had greater opportunities to engage in social projects (Chi, 2011).

Conclusion

Post-Museum's Rowell Road period (2007–2011) developed and provided a co-constitutional space for the arts and civil society, where artists, activists and civic actors met in a space where they could push boundaries and be free to engage with social and political issues which state-led institutions would often discourage. In a way, Post-Museum functioned as a 'micro-utopia' that proposed an alternative vision of Singapore's 'renaissance' in the hope of realising a more open and liberal Singapore. Although the Rowell Road Period lasted for only four years, it was one of the few artists-initiated spaces that opened its doors and invited diversity and progressive thought.

However, independent art spaces often face the problems of lack of financial resources compared to state-led initiatives and this results in the short life-cycle of these initiatives. Post-Museum ended its space on Rowell Road in 2011, due largely to the lack of financial resources. In Singapore, there is a trend where art groups like TAV operate without a physical space to circumvent financial constraints. Post-Museum continues to operate nomadically to organise, curate, research and collaborate with a network of social actors and cultural workers. This nomadic model has resulted in a different practice from its earlier Rowell Road period. The Rowell Road period is an important period for Post-Museum as they are the formative years, which allowed Post-Museum to define its social practice in its current nomadic model. Though there were many successes and failures in the Rowell Road period, this chapter also argues that independent art spaces can play a critical role in providing a space for people to imagine and create new visions in Singapore's 'renaissance'.

In way of a conclusion, we would like to raise Tham's finding that issues of financial sustainability often overshadow issues of succession planning in ensuring the continued relevance of such independent art spaces. Tham argued that the government should allocate more resources and research towards independent art spaces as they develop what cultural economists call an intangible cultural capital, and in that way, play a more critical role in Singapore's context of the development of a contemporary art scene and market. We agree with Tham's views and would add that financial support for independent art spaces should be coupled with respect for autonomy and flexibility of deciding what to focus on and how to work. In other words, government funding should not be a way to subsume community-led independent art spaces into government-led development.

The vision of community-led independent art spaces like the Post-Museum in its Rowell Road Period are markedly different from the government's vision of Singapore's 'renaissance', where the definition and limited scope for the arts seems to be market-driven, decorative and non-disruptive. However, this should not be seen merely as a 'position of resistance', but as providing a valuable, even critical, contribution towards Singapore's 'renaissance'. Therefore community-led independent art spaces should be allocated more government resources.

Notes

1 Kampung is a Malay word for rural village.
2 Artist Daniel Buren, in his seminal text on the function of the studio where the studio is a site of production, said that the studio was important in providing the space for artists to develop the first 'frame' (Buren, 1979).
3 p-10 was an independent curatorial team based in Singapore from 2004 to 2008. Its focus was on the development of artwork and areas surrounding the practice of art. The co-founders were Lee Sze-Chin, Lim Kok Boon, Charles Lim, Jennifer Teo and Woon Tien Wei. Subsequently, Cheong Kah Kit joined as a member of p-10.
4 Little India is Singapore's foremost Indian enclave under the British Colonial adminis-tration. Under PAP's policy of racial harmony, though people from ethnic Indian groups do not live solely in this area, it is maintained as a cultural heritage area where many of the old commercial or cottage industries still remain. Today, there are a mix of commer-cial and residential estates in the area. With the influx of South Asian migrant workers, Little India has become a popular area for this community on their rest days. Many view Little India to be very 'un-Singapore' as it is the most messy, dirty and chaotic part of Singapore (Tee, 2010).
5 Karung Guni shop is a shop which deals with secondhand items.
6 A shop-house is a kind of architectural building type that is commonly seen in Southeast Asia. Shop-houses are usually two or three stories high where the shop is located on the ground floor and upper floors are residences. For more information see en.wikipedia. org/wiki/Shophouse.
7 In 1989, an advisory council was formed to study the arts and culture in Singapore and as a result, it then published the Advisory Council on Culture and the Arts, which presented recommendations designed to make Singapore a culturally vibrant society by the turn of the century. This report was a predecessor to the Renaissance City Plan and marks an ideological shift in the government's perception of the role that arts and cul-ture would play in Singapore's future ('Report of the Advisory Council and the Arts', 1989).
8 Speaker's Corner is fashioned after Speaker's Corner in Hyde Park, London. Although Singaporeans are required to submit the identity details to the police stationed there, political expression is allowed here and no license is required; non-Singaporeans are not allowed to speak there.
9 Government-led cultural development refers to initiatives to develop arts and culture which are initiated by the government and managed by statutory boards or state-funded agencies.
10 Community-led cultural development refers to initiatives to develop arts and culture which are initiated by the arts community based on their own resources and managed independently of the government.

11 Food #03 is a development from Food #02, an artwork/performance Woon Tien Wei created in 1999/2000 while he was studying in the UK. It was based on an appropriation of American artist Gordon Matta-Clark's restaurant called Food, which he opened in Soho, NY, in 1971. In the midst of planning for Post-Museum, there were plans to create a food and beverage area where visitors and exciting creative people can hang out. Instead of renting the space to another business, Woon decided to create Food #03, which is a deli-bar, social-enterprise, source of revenue for Post-Museum, hang-out venue for like-minded individuals to mingle, platform for new exciting ideas and work of art.

12 The known illegal trades include sale of pornographic DVDs, 'smuggled' cigarettes, sex enhancement drugs, gambling dens and brothels.

13 Unlike most government-led food programmes, The Soup Kitchen Project does not do means testing. Volunteers identify locations which house the poorer populations. Volunteers go door to door and asked if the residents needed free dinner every Monday. For more information on The Soup Kitchen Project, see www.facebook.com/thesoup-kitchenprojectsg.

14 The Substation, opened in 1990, was Singapore's first art centre and it played a crucial role in developing the arts as it provided the much-needed platform for artists to showcase and develop their work.

15 As part of the Labour Day celebrations, Migrant Voices presented works done largely by work-permit holders (foreign workers) through past workshops, including paintings, drawings, writing and photographs.

16 An exhibition to provide an educational platform regarding the work of NGOs and other cause-related organisations in Singapore and the region. Targeted for the Christmas season, the audience, through purchasing products which satisfy their needs and wants, contribute to the various causes. For more info see post-museum.org/shopperhero/index.html.

17 Post-Museum in partnership with Singapore Queer-Straight Alliance, produced an exhibition as an event that was part of IndigNation 2009. For more info see www.post-museum.org/allyouneedislove/submit.html.

18 This project looks at the past, present and future of civil society in Singapore. Featuring an exhibition, talks and other events, this series will be a foray into the challenges and opportunities of organising for real change in Singapore. For more info see http://www.post-museum.org/funds/acc.html.

19 In 2010, the newly formed Socialist Front held its first press conference in the Show Room on 29 October.

20 In 2011, MARUAH and The Online Citizen held the Post-Elections Public Forum on 15 May one week after the general election.

21 Chi used geographer Robert Sack's notion of a 'good' place to describe co-constituted creative spaces/communities, which serves both the arts and civic spaces (Sack, 2003). Places which allow people to see the world clearly and as deeply as possible and that promote variety and complexity are 'good' places that expand our knowledge of the good and hence our ability to do good. Chi observed the utopian vision of Post-Museum where it functions as creative space that aims to facilitate diverse and complex projects which could potentially expand knowledge about contemporary issues and consequently progress beyond the Post-Museum's space fits Sack's description of 'good' places.

References

'About Post-Museum' (2009), retrieved 29 May 2010, from post-museum.org/about.html

'The Artists Village: About us' (2005), retrieved 28 June 2009 from http://tav.org.sg/index.html.

Benjamin, P. (2008) 'The do gooders', retrieved 1 December 2015, from sg.asia-city.com/city-living/article/do-gooders.

Buren, D. (1979) 'The function of the studio'. *October*, *10*, 51–58.

Chi, H. (2011) 'Can space speak? Independence, creativity and social action in Singapore: A case study of Post-Museum'. (BA Bachelor of Social Sciences Thesis), National University of Singapore, Singapore.

Chua, B.-H. (2008) 'Culture and the arts: Intrusion in political space'. In K. F. Lian and C.-K. Tong (eds.), *Social Policy in Post-Industrial Singapore*, Boston: Leiden, 225–245.

Enwezor, O. (2006) 'The artist as producer in times of crisis'. In S. McQuire and N. Papastergiadis (eds.), *Empires, Ruins + Networks: The Transcultural Agenda*, Melbourne: Rivers Oram Press, 11–51.

Hamnett, C. (1995) 'Controlling space: Global cities'. In J. Allen and C. Hamnett (eds.), *A shrinking world?* Oxford, UK: Oxford University Press.

Ho, V. (2013) 'Public morality in sex spaces'. [Web Journal]. *S/PORES*.

Koh, G. and Ooi, G. L. (2004) 'Relationship between state and civil society in Singapore: Clarifying the concepts, assessing the ground'. In H. G. Lee (ed.), *Civil Society in Southeast Asia*, Singapore: Institute of Southeast Asian Studies, 167–197.

Kong, L. (2000) 'Cultural policy in Singapore: Negotiating economic and socio-cultural agendas'. *Geoforum*, 31, 409–424.

Kuo, E. (2000) 'Epilogue'. In G. Koh and G. L. Ooi (eds.), *State-Society Relations in Singapore*, Singapore: Institute of Policy Studies.

Kwok, K. C. (1996) *Channels & Confluences: A History of Singapore Art*, Singapore: Singapore Art Museum.

Kwok, K. W. (2009) 'Introduction: Locating and positioning the Artists Village in Singapore and beyond'. In K. W. Kwok and W. Lee (eds.), *The Artists Village: 20 Years On*, Singapore: Singapore Art Museum, 1–3.

Lee, T. (2005) 'Gestural politics: Civil society in "new" Singapore'. *Sojurn: Journal of Social Issues in Southeast Asia*, *20* (2), 132–154.

Lee, T. (2007) 'Towards a "new equilibrium": The economics and politics of the creative industries in Singapore'. *Copenhagen Journal of Asian Studies*, 24(1), 55–71.

Lim, P. L. (2009) Casino Control Act. *Singapore Infopedia*, retrieved 23 August 2016.

MICA (Ministry of Information Communications and the Arts). (2008). *Renaissance City Plan III Public Report Unveils Details of Singapore's Arts and Cultural Master Plan for 2008–2015*. Singapore: MICA Retrieved from app.mica.gov.sg/Default.aspx?tabid= 36&ctl=Details&mid=539&ItemID=931.

MITA (Ministry of Information and the Arts) (2000) Renaissane City Report 'Culture and the Arts in Renaissance Singapore' (Vol. 2006), Singapore: Ministry of Information, Communications and the Arts.

Ooi, C.-S. (2010). 'Political pragmatism and the creative economy: Singapore as a City for the Arts'. *International Journal of Cultural Policy*, 16(4), 403–417.

Report of the Advisory Council and the Arts (1989), Singapore.

Robinson, J. (2006) *World Cities, or a World of Ordinary Cities? Ordinary Cities: Between Modernity and Development*, Abingdon, UK: Routledge.

Sack, R. (2003) *A Geographical Guide to the Real and the Good*. London: Routledge.

Sasitharan, T. (1990, 24 February) 'A power house of dreams', *The Straits Times*.

Sassen, S. (1991) *The Global City: New York, London, Tokyo*. Princeton, CN: Princeton University Press.

Storer, R. (2007) 'Making space: Historical contexts of contemporary art in Singapore'. In R. Storer, E. Tan and G. Nadarajan (eds.), *Contemporary Art in Singapore*, Singapore: Institute of Contemporary Arts Singapore, 9–18.

Tan, E. (2007) 'Singapore: Too contemporary for art'. In R. Storer, E. Tan and G. Nadarajan (eds.), *Contemporary Art in Singapore*, Singapore: Institute of Contemporary Arts Singapore, 24–29.

Tan, K. P. (2007) 'Optimists, pessimists, and strategists'. In K. P. Tan (ed.), *Renaissance Singapore?Economy, Culture, and Politics*, Singapore: NUS Press, 253–269.

Tee, E. (2010) 'A walking guide to Singapore's Little India'. *CNNGo* retrieved 25 May 2010, from www.cnngo.com/singapore/shop/walking-guide-singapores-little-india-119396.

Tham, Y. (2009) 'From fringe to market: The role of independent art spaces in the development of Singapore's contemporary art market'. (Master's Thesis), Sotheby's Institute of Art/University of Manchester, Singapore.

Turner, C. and Barclay, G. S. J. (2005) 'Singapore: A case study'. In C. Turner (ed.), *Art and Social Change: Contemporary Art in Asia and the Pacific*, Canberra: Pandnus Books, 267–277.

Urban Redevelopment Authority. (2003). Master Plan 2003, retrieved 23 December 2015, from   www.ura.gov.sg/uramaps/?config=config_preopen.xml&preopen=Master%20Plan%202003.

Zukin, S. (1995) *The Cultures of Cities*, Oxford, UK: Blackwell.

3 Art/movement as a public platform

Artistic creations in the sunflower movement and the umbrella movement[1]

Pei-yi Lu and Phoebe Wong

In the spring of 2014, the Sunflower Student Movement in Taipei protested against the Cross-Strait Service-Trade Agreement (CSSTA), while in the fall, the pro-democracy Umbrella Movement in Hong Kong drew the world's attention to the young protesters' non-violent resistance, as well as their determination and persistence. These two protest-cum-occupations triggered critical discussions about democracy, identity and the future, while the large quantity and wide variety of objects, images and performances created during the respective demonstrations were unprecedented. How should they be considered from the perspective of contemporary art practice?

Under the modernist paradigm, art is seen as autonomous, and thus art is expected to keep its distance from politics to maintain its freedom of creation. At the far end of the spectrum, however, art is often used to serve political purposes during certain periods in communist countries. With the recent re-emergence of art activism, a dialectical debate on the politicization of aesthetics and the aestheticization of politics is prevalent in the contemporary art field (Groys 2014).

In recent times, the overlapping practices of art and activism have been addressed in a number of large-scale survey exhibitions (accompanied by publications), such as: 'Living as Form: Socially Engaged Art from 1991–2011', conceived by Creative Time in New York in 2012; the seventh Berlin Biennial 'Forget Fear' in 2012; 'global aCtIVISm' presented by ZKM | Center for Art and Media, Karlsruhe, Germany in 2013–14; and 'Disobedient Objects', held at the V&A Museum in London, and which was staged from the perspective of design.[2]

These exhibitions have shed a new light on rethinking the role of art within social movements, neither seen as a tool nor as something useless. As Peter Weibel, the curator of 'global aCtIVISm' writes, 'for some years now, a new form of world-wide activism driven by citizens (lat. civic) has been in evidence, as the word CIVIS highlighted in aCtIVISm emphasizes. It is a movement spawned by globalization, technological developments and the expansion of art' (Weibel 2015: 23). This 'expansion of art' — its manifestation, its directions, and its depth and breadth, deserve some closer attention. In the post-exhibition publication, *Global Activism: Art and Conflict in the 21st Century*, Weibel further asserts that the proliferation of 'artistic performative practices' appeared as 'performative democracy' (to borrow from Elzbieta Matynia) and as 'a new form of public art' (Weibel 2015: 59).

In this light, from the perspective of contemporary art, this chapter examines the two movements through the following questions: In today's highly connected world, what are the similarities and differences between the Sunflower and Umbrella Movements within and beyond global/local resistances? What role can art play in these movements? How should one view these creative expressions[3] which are made and used during protests? What is the potential of the visual aspects of these movements to appeal to political sensibilities? Do they serve as propaganda or do they effect political or social change?

The geopolitics

The context of the Sunflower Movement and the Umbrella Movement

Both the Sunflower Movement and the Umbrella Movement took place in 2014, but what were the global/local contexts of these two movements? After the 2008 global financial crisis, many demonstrations occurred for a variety of reasons, including the Arab Spring in 2010, striving for democracy, and the worldwide 'Occupy' movement that started off from the 'Occupy Wall Street' demonstration, protesting against corporate greed and social and economic inequalities. It is worth noting that Hong Kong's 2011–12 'Occupy Central' campaign (camping out in the plaza beneath the HSBC headquarters for nearly eleven months) remains one of the lengthiest 'Occupy' movements in the world. To a certain degree, these cases have encouraged and inspired political and social movements going forward in terms of their concepts, forms and methods. The discussion here recognizes that global activism sets the backdrop for the occurrences of the movements in question, while the regional geopolitics in relation to China have determined the course of the two events as civil disobedience movements.

The Sunflower Movement: Against CSSTA

The Sunflower (student) Movement (from 18 March to 10 April 2014) was aimed at protesting against the illegal passing of the CSSTA by the ruling party, the Kuomintang (KMT). This unlawful practice was regarded as an 'under-the-table' agreement, or in the protestors' terms, a 'black box' operation. The passing of the CSSTA reflected a crisis of democracy, and was seen as an act of Chinese incursion into Taiwanese sovereignty.

The lifting of martial law in Taiwan in 1987 was a pivotal milestone in the development of democracy for the country, and this was followed by the process of 'Taiwanization' in many social-political-cultural aspects. Since the 1990s, following the Wild Lily Student Movement,[4] Taiwan has gone through a series of democratic reforms, not least of which was the first direct presidential election in 1996. The first change of government happened in 2000 when the presidential candidate of the Democratic Progressive Party (DPP), Chen Shui-bian, took office. The DPP was in power for eight years, but the last couple of years were marred by corruption scandals surrounding President Chen. In 2008, the KMT returned to power as Ma Ying-jeou became president. Its pro-China stance sparked controversy, and China took advantage of the situation by putting business over politics, and hoping to achieve its goal of unity through economic tactics (以商圍政, 以經促統) (Wu 2014: 37). Since then, many social movements have occurred,[5] culminating in the Sunflower Movement in the spring of 2014.

The first direct action of the Sunflower Movement was in the form of an occupation, which took place on the evening of 18 March 2014, when both students and citizens broke into the Legislative Yuan (the state's parliament). Subsequently, surrounding streets were also occupied to prevent the police forces from expelling students from the parliament buildings. On the second day, sunflowers were sent to the parliament to encourage the students, which subsequently gave the name to this movement. Several important events occurred, as follows. On 24 March, a group of protesters occupied the Executive Yuan and was forced to leave by the police ten hours later. On 29 March, an advertisement, 'Democracy art 4 pm,' was published on the cover of the *New York Times* as a criticism against the government's violence. Also, on 30 March, around 500,000 people rallied to give voice to the accumulated anger and disillusionment of the people. The whole event ended on 10 April after a lengthy discussion between the student protesters and non-government organisations (NGOs). During the twenty-four days of occupation, the streets were turned into living spaces veiled with a festive atmosphere. Creative expressions were largely produced and installed on site.

Although this movement was triggered by an economic issue, it has since generated more significant considerations about 'nation' (the politics) and 'community' (civil society). As a consequence of this movement and the continuous efforts made, a new type of political imagination has been created. Following the actions of the movement, the KMT suffered heavy setbacks in the local elections, which took place towards the end of 2014, and the subsequent loss of the 2016 presidential and parliamentary elections. The Sunflower Movement also signalled the establishment of an independent mindset among younger generation Taiwanese. Relations between Taiwan and China have since entered a new phase, which may have an influence on the balance of international power between China, America and Japan in the Pacific Region.

The Umbrella Movement: 'I want genuine universal suffrage'

The Umbrella Movement (from 28 September to 15 December 2014) was the largest civic movement in Hong Kong since the territory's support of the 1989 Tiananmen student movement and the 1 July rally in 2003. Similar to the case of the Sunflower Movement, the increasing political encroachment of China triggered the occurrence of this pro-democracy movement.

As part of the historic handover in 1997, the political destiny of Hong Kong was ostensibly settled as a Special Administrative Region (SAR) of China under the inventive 'One Country, Two Systems' formula. However, the problems of post-coloniality (from British rule) and neo-colonialization (by China) have been haunting the city ever since (Erni 2001). The rapid social and economic integration with China, and Beijing's increasing interference in Hong Kong's affairs, have eclipsed the autonomous status quo of the city, and have been bringing resistance and hostility against both the Chinese government and Chinese people from the mainland. As some political scholars have observed, the '[d]eteriorating human rights records and tightened political control from Beijing since 2008 aggravated the "anti-China' sentiments", and radical politics have emerged in every aspect of Hong Kong society in the past decade' (Ma 2015; Cheng 2014). Indeed, a part of this 'anti-China' sentiment showed as a process of 'unfriending' China, when one considers the ups and downs of the China-Hong Kong relationship over recent decades. Indeed, back in the mid-1990s, with the economic prosperity in the region, the brain drain from Hong Kong after the Tiananmen crackdown was abated, while amidst the confidence crisis it faced in returning to China, Hong Kong embraced the idea that with its international outlook, its openness and its more advanced judicial system, it could make valuable contributions to China's development.

Over the past three decades, Hong Kong has been striving for democracy – since the commencement of negotiations between China and Britain regarding the future of Hong Kong, which resulted in the signing of the Sino-British Joint Declaration in 1984. The 2014 constitutional reform was about the implementation of a promised universal suffrage for the Hong Kong SAR chief executive in 2017. The Hong Kong people have demanded the provision of genuine and fair elections in its design. In opposition to the decision of Beijing's Standing Committee of the National People's Congress (NPCSC) of 31 August 2014 – which imposed restrictions on the nomination system of the election – a city-wide class boycott campaign led by students' organizations such as the Hong Kong Federation of Students and Scholarism, was carried out in late September, and was followed by a proclamation of the start of the Occupy Central with Love and Peace (OCLP).[6] The occupation soon spread to a number of sites in the city, including Admiralty, Mong Kok and Causeway Bay, and lasted for seventy-nine days.

In summary, the Sunflower and Umbrella Movements should be considered as part of the global citizen movement, but they are deeply rooted in the 'anti-China sentiments' that prevail in Taiwan and Hong Kong. The concerns of the two movements are strikingly similar: 'Taiwan's activists yelled, "It's our country, we will save it; it's our future, we will decide it". Meanwhile, Hong Kong activists cried, "I protect my city; we decide our own fate"' (Lin 2014). Furthermore, the controversial phrase 'Today's Hong Kong, Tomorrow's Taiwan' evinces a sense of their shared destiny.

Visual politics in the Sunflower Movement and the Umbrella Movement

A global citizen movement in the twenty-first century is characterised by a bottom-up, participatory model, which involves the using of the Internet (especially through social media), mobile phones and citizens' own bodies, and which results in the formation of a civil society that 'reigns in a spirit of hope, inclusiveness, and improvisational genius' as summarized precisely by San Francisco-based essayist Rebecca Solnit (2014: 28; Chang 2014; Lee 2015).[7]

With a clear goal and under the need for affectivity, flexibility and imminent progress during a social movement, the occupied sites could be seen as battle-fields. In such a situation, what is the meaning of these sites, where in Taiwan, it is linked to the parliament buildings and the surrounding streets, while in Hong Kong, it takes place in a variety of urban landscapes? How should any collective making be viewed where it may have been produced by a citizen-as-artist, or how should any artistic creations by artist-as-citizens be considered? In the next section, a number of aspects revealing the visual politics of the movements will be discussed, as follows:

1   spatial arrangement,
2   symbol making,
3   collective making,
4   the debate about artist-as-citizens' involvement, and
5   any resultant archives, exhibitions and publications.

The Sunflower Movement: Creative creations

Parliament as a living place

Place making and spatial rearrangement forms are the first layer of visual politics. The occupation by the Sunflower Movement was in a concentric circle mode, with the parliament in the centre and the surrounding streets at the edge. Inside the parliament, the space was adapted as a resident base with well-defined functions: a huge media section with umpteen cameras in the middle, lawyers on the right, art propaganda on the left, and a resources centre and sleeping area at the back. It is interesting to note that 'speaking out' is the most crucial element of the spatial arrangement. In addition, as a result of the twenty-four-hour live broadcast, it was similar to a reality television show with links to symbols of transparency/visibility.

Outside the parliament, people also turned the street into a living space. Living on the street prompts reflection, raising questions such as what is the nature of life and what is needed for a better quality of life? Streets were turned into a public square, serving as a meeting point for conversations, discussions and the exchange of ideas. This movement made the streets into a 'place' with co-living practice, filling emotion and having relations during the period of occupation. It made daily life into a form of resistance. In this sense, occupation is a kind of 'pre-figurative politics' (Graeber 2014) that practices a possible future and creates more space for the imagination. Also, in the case of the Sunflower Movement, 'the various events happened on the streets that made the tense of inside and outside transform into a dialectical relation, rather than developing into violence' (Minato 2015: 119).

Collective making

Taking on the name 'Sunflower Movement' happened by mere chance. The name was adopted after sunflowers were sent to the parliament buildings as a gesture of support the day after the occupation had taken place. Apart from being an eye-catching icon, in Taiwan the sunflower alludes to a bright and hopeful future. Also, using a flower as a name for student movements has become something of a tradition in the recent history of Taiwan, including the Wild Lily Student Movement (1990) and the Wild Strawberry Student Movement (2008). However, the symbol of a sunflower has remained on the surface and has not been taken deeply.

As this movement was a revolt against the illegal passing of the CSSTA, which was seen as an 'under-the-table' agreement, the metaphor of a 'black box' became the crucial image of this event. The students' occupation of the parliament was an action designed to return the site to its original function as a place of speaking-manifesting opinions. The Japanese photographer-cum-curator Chihiro Minato emphasizes the significance of 'eyesight', which is usually a major asset of artists. On the one hand, it means that students can 'see through' the existence of a black box, the invalidity of representative politics (parliamentary democracy), while on the other hand, it also refers to the 'vision' they want for the future (Minato 2015: 98). In this vein, visibility could be seen as a key concept for this movement. Indeed, when the protests first began, the mass media regarded the students as 'violent people', which made the other participants try to cast off this derogatory name and maintain the image of the 'good citizen'. Several binary oppositions emerged: the black box (invisible, dark) versus transparency, violent people versus good citizens and evil versus justice. These metaphors were presented in the banners, posters, graffiti and objects of disobedience. They could be seen as a 'representation' of visual politics.

Collective making was commonplace, especially on the streets outside the parliament buildings. As 'the truth must be concrete', 'through action and making an effort', people can share feelings and emotions. They need to make objects, write texts and create sounds. The process of embodiment is the subject of the politics (Minato 2015: 88). Based on this idea, collective making could be linked to the presentation of the politics.

The visual culture scholar Chang Shih-lun argues that a participant actively presenting images made by themselves is a distinctive feature of the Sunflower Movement (Chang 2014: 26). In the past, social movements relied on mainstream media to report, and that is the 'representation' of the reality. This time, the participants 'presented' their views and appeals via images to reveal something that had been hidden or placed undercover. These images claim that 'I am here', 'I see it', 'I participate' and, therefore, 'I am a witness and part of this movement'. Some projects on site belong to this type. For example, 'Witness Democracy' was about people taking protest banners on site as a performance; 'Art People in the Parliament' was about taking a frame to reveal the reality inside parliament. A further project asked for foreigners to hold posters in support of the movement. In these projects, features include the participation of normal people, the emphasizing of their sensibilities and the images captured by the omnipresent camera.

The feature of an 'affective turn'[8] (Chang 2014: 26) can also be observed in the objects made in the movement. Here, the song 'Island's Sunrise' is a good example. A singer-songwriter wrote the song after the violent occupation of the

Executive Yuan with the idea that 'we need some gentle and soft power.' Within a few days, this song became a part of the collective memory of this movement. The song is about a young man who faced injustice and who was prepared to fight for his own rights, so he said sorry and goodbye to both his lover and his mother. This song also adopts the metaphor of darkness/brightness, referring to the end of the night and the oncoming sunrise. On many occasions, this song deeply touched the hearts of the protesters and encouraged them to stay until victory was achieved. Minato pointed out how the idea of 'care' could be considered as a possible reason for this non-violent event (Minato 2015: 88), not only among the protesters but also emerged between the protesters and the police.

The debate about artist-as-citizens' involvement

According to the survey 'Who joined the Sunflower Movement?' (Chen 2014), 9.3 per cent of student protestors came from the fine art discipline. If the demography is expanded to include Design and Architecture/Urban Planning, the figure rises to 16 per cent. This may explain why the quality and quantity of creations are considered more than previous social movements. Some of the creations were made collectively by ordinary people, some would have been artist-led participation, some were art projects created by anonymous artists, and only a very few were identified as 'artwork' by the artist who showed the artist status. In this vein, artists-as-citizens involved themselves in various ways.

Furthermore, in its very early stages, the Sunflower Movement was nominated as a year great exhibition in the Taishin Art Award by the scholar and jury member Chang Hsiao-hung. This nomination was controversial since the idea of activism has previously not been included in the field of art. This nomination triggered a debate and raised the question of how artistic creations should be viewed within the social movement.

Two cases outlined below will illustrate the issues of artists creating works in the context of an occupation. 'Occupy the 138th hour', by the artist Chen Ching-yua,[9] who was with the first group of students entering the parliament and who played various roles in this movement, including as a protester, an artist, a recorder, a producer, and so on, was made on site as the progress of making oil realism painting. This painting caught up a shot of the 138 hours presented the fifth day of occupy. During the occupation, as Chen created this painting in the middle of the parliament member seats for a few days, Chen Ching-yua attracted much attention and criticism, especially from the mass media. The comments included assertions such as 'live painting is not new, rather [it is] returning to the age of Modernism: dominated by painting, full of grand discourse, and the worship of icons' and 'the key should be the performance of artist's status' (Wong Ying-dah 2014). The feeling of worship was strongly built when the end of occupy, artists Chen held this painting as a ritual to walk out the Parliament. He also refused to allow his painting to be included in the public archive project held by Academia Sinica. He claimed his painting should be in his agent's foundation that raised more criticism from the contemporary art field.

Another video piece, 'Occupy the 516th hour' by artist Yuan Goang-ming, recorded the final two days before the protestors withdrew from the parliament. In the beginning, the purpose of Yuan's filming was to record the final moments of the occupation to provide visual material for the English version MV of 'Island's Sunrise'. For Yuan, it might fall into 'illustration-based narrative'. He decided to use his usual artistic methods to create this work. In the film, all the details inside the parliament were presented, like scanning the scenes, until returning to the original situation. The sound-track of this piece derived from the national anthem which was slowed down and reversed to create an atmosphere which was both haunting and sacrificial.

Both this piece and Chen's paintings were later shown in the exhibition entitled 'Asian Anarchy Alliance'[10] in an art museum. This form of institutionalization makes the relationship between art and social movement more complex. How should these works be considered, either as products of a collective effort or belonging to an individual artist's intelligence? On the one hand, an artistic work could be used as an example to refer to a particular event. Once it is shown, discussion may be re-opened as part of a public forum. Thus, it appears to be back in the public domain. On the other hand, without the Sunflower Movement, these two particular works would never have been created. The works should remain as a part of the event. However, ultimately, ownership remains in the private sector rather than the public sector.

If these two pieces of work were recognized as artworks, tagged with artists' names, and shown in museums, how should we view the objects that were made from functional needs with a collective effort by nameless individuals? The Hill of Chairs (as a makeshift barricade) is an exceptional case to illustrate the concept of collective making. The debate was around the issues of how we see art/movement. On the first night, when students occupied the parliament, they heaped up chairs found inside the building to block one side entrance to prevent the police from raiding the premises. The scholar Chang Hsiao-hung considers The Hill of Chairs as a new form of art installation (Chang Hsiao-hung 2014), whereas another critic, Wong Ying-dah, argues that 'The Hill of Chair is an "object of resistance". That is, it is a manifestation of the successful mobilization of the politics, of the masses (i.e. the people), of the technical know-how, and of emotions. It reflects the complexity of labour relations and bodily experiences that exist between people, movement, and material' (Wong Ying-dah 2014). This view is shared by Chang, and as she wrote a year later, 'art-movement should be considered from its intensity rather than a form of representation' (Chang Hsiao-hung 2015a). Looking from the perspective of photography and documentation, another scholar, Gong Jow-jiun, pointed out that through the redeployment of sensibility, the texture of objects, such as The Hill of Chairs, could be caught by photographers as 'haptic visuality' which should be valuable documents for the future (Gong 2015).

In the case of the Sunflower Movement, occupying the parliament was a means of suspending the existing situation to create an exceptional moment which could offer the potential to alter the reality. The Sunflower Movement showed its resiliency both in its creativity and in its thinking and practicing. This was the first time that such a high percentage of art practitioners were involved in such a move-ment. In the role of artist-as-citizen, they took actions through their abilities and sensibilities of art which made this movement distinctive. However, the debates outlined below show that the perspectives of art and social movement have shifted to the reconsideration of the possibilities of Art/Movement. In addition, the soft power of art may have assisted this movement in moving towards a state of non-violence and the potential to change the reality.

The Umbrella Movement: Collective making

Separate occupied sites

The Umbrella Movement happened mainly in three sites within Hong Kong. In Admiralty, the occupation congregated on a section of Harcourt Road adja-cent to the SAR government headquarters and the LegCo building, from where it spread east to Wan Chai and west to the fringe of Central. Harcourt Road, a major thoroughfare with four lanes of traffic on each side, was transformed from a 'non-place' – a term defined by anthropologist Marc Augé (1995) to refer to an anthropological space of transience, and a motorway is a typical non-place – into a place charged with deep emotions and meanings within a couple of weeks after the occupation took place. This occupied zone was nicknamed 'Harcourt Village' by the protestors; the deliberate choice of the word 'village' contrasts with the metropolitan outlook of Hong Kong as a global, neo-liberal economic entity.

Protestors also seized the busiest junctures in the Mong Kok and Causeway Bay districts, both of which are sites which combine business, shopping and resi-dential areas. The disparate atmospheres found in the Admiralty and Mong Kok areas suggested a de-centring or even disunity and dissidence within this move-ment, with a self-disciplined, well-organised Admiralty occupied zone compared to a chaotic and gangster-harassed Mong Kok. Indeed, according to one observer, 'some commentators […] endowed human qualities upon the three occupying zones: Admiralty is "chic"; Mong Kok is "masculine"; Causeway Bay is "kawaii (cute)"' (Lau, Kwong and Siu 2015: 148).

Umbrellas, sticky notes, Lion Rock spirit

The name 'Umbrella Revolution' was coined by Adam Cotton on Twitter on 29 September, with reference to the umbrellas used as a form of defence against pepper spray and tear gas.[11] The name quickly gained widespread acceptance. Soon after, *Time* magazine, an Asia edition dated 13 October 2014, used a pow-erful image on its cover, featuring protestors with their umbrellas bathed in a mist of tear gas, following which the motif of the umbrella was widely used and it instantly became the icon of the movement. If pepper sprays and tear gas are seen as objects of de-demonstration, then the umbrellas are objects of de-de-demonstration. The umbrella was transformed from the functional – for protec-tion – to the symbolic – a symbol of resistance. Riding on its inherent graphical appeal, umbrella images can be conveniently presented or transformed into a large variety of forms and motifs. The rich productions and reproductions of the motifs, and their wide dissemination, collectively have sustained the synergies of the movement. Incidentally, there is a strong graphic resonance between the two movements in question in this paper. As mentioned above, there is a convention, although sometimes arbitrarily, for Taiwan to name student movements after flowers, and the graphic resemblance between an umbrella and a flower draws the two movements together in a 'visual bond' (in Dziga Vertov's sense), thus creating a closer link. One key slogan of the Umbrella Movement, *pin dei hoi faa* (遍地開花) literally meaning 'blossom everywhere', metaphorically points towards the messages of resistance flourishing on a large scale.

Unlike the case of the Sunflower Movement, there are no statistics to mark out the protestors as artists or otherwise in the Umbrella Movement; however, it is clear that many artists became involved in view of the emergence of art activism and artist-as-citizens in the past decade in the city. Indeed, one of the first pieces of art to appear at the Admiralty site was a quilt made of torn umbrellas collected from the site by art students from the Academy of Visual Arts at the Hong Kong Baptist University. The quilt, forming a canopy, was hung between the twin footbridges that run across Harcourt Road from Admiralty MTR station to the government headquarters. With many huge banners also hanging down from the footbridges, and the emergence of the Lennon Wall, this area soon became the centre of the Admiralty occupied zone – essentially creating an informal 'big stage' for public forums, and for announcements concerning the movement.

The Lennon Wall, Hong Kong, named after the Lennon Wall in Prague, in the Admiralty protest zone is emblematic. The Wall, initiated by a social worker, has become a visual statement of solidarity and the collective voices of the movement. The wall is basically composed of sticky, post-it notes, with a message on each note. People wrote down their pleas, demands, views, hopes, feelings, or words of support on the notes. A single note being so small in size perhaps denotes a faint, individual voice. However, when many notes are placed together, such small voices are amplified, and turned into something powerful. When this sense of collectivity is infused with the ideological rejection of a 'big stage' and a single, centralized occupied zone, it constitutes a key attribute of the Umbrella Movement whereby 'no leadership' can be seen as a manifestation of a general dismissal of representative politics.

A gigantic yellow banner, six meters wide by twenty-eight meters long, with a printed Chinese slogan, 'I want genuine universal suffrage', appeared on the Lion Rock, located in central Kowloon, on the morning of 23 October 2014, but it was taken down by the authorities the next day. The appearance of this banner marked a 'significant and sentimental moment' (Cheung 2016) in the occupy movement. People immediately felt connected because:

1　as a spectacle, the public was amazed by the well-executed hanging of a banner of such monumental size on the rugged ridge of the Lion Rock by a group of anonymous rock climbers;

2   the action offered an uplifting moment when the movement came to a stand-still after almost a month, and neither the government nor the protestors could envision a solution;

3   it worked as a renewal of the 'Lion Rock spirit',[12] that infused the action with a new found political meaning (Cheung 2016).

Politics of memory: Documenting, archiving, exhibiting

One peculiar aspect of the Umbrella Movement is that there has been a strong archive and documentary impulse. It is worth noting that the notion of archiving came into place even in the early stages of the movement as it unfolded. The proactive documentation initiatives from below through certain protestors' setting up of archives or archival projects, producing documentaries and publishing, contrast with the government's non-responsive attitude. Concerned individuals, such as Dr Simon Chu Fook-keung, the former director of the Government Records Service of Hong Kong, and a long-time proponent of establishing the archives law in Hong Kong to supervise the action of the government to protect valuable government records and dossiers, expressed concerns that the SAR government might destroy the Occupy Movement's records and documents as there was no implementation of archives law in place in the territory.

One distinctive case is the Umbrella Movement Visual Archives & Research Collective (UMVARC), spearheaded by urbanist and cultural researcher Sampson Wong (see his contribution to this volume). This documentation initiative began as early as a week after the Umbrella Movement had begun. A group of volunteers (up to 150 at one point) systematically recorded – photographing and logging – the diverse expressions of citizens' participation in a controversial, political environment (Wong 2016 and this volume).

Initiated by a small art organization called The Library by soundpocket, 'The Umbrella Movement – Field Recoding Investigative Project' was the first of its kind to collect sonic material from a social movement. As soundpocket's in-house editor and researcher, Law Yuk-mui, explained,

> it is about how one experiences sound, observes sound, understands sound and responds to sound through recording a bona fide social movement as well as using sound as a tool of resistance. The investigation takes the Umbrella Movement as a case study, where artists investigate sound and its associations or interpretations through their actual participation and experience.[13]

For this five-month project (January to May 2015), a total of thirty-five sound clips from eight artists were collected, and a CD album titled *DAY AFTER [2014.9.29 – 12.12]* was released in 2015. Apart from these archival initiatives, 2015 also saw a large number of exhibitions, publications and documentaries related to the Umbrella Movement. It is estimated there were approximately a dozen exhibitions, over twenty documentaries, both long and short (of which, two were three-hour feature-length documentaries), and over fifty publications including text-based, photography books, art books and the like. All in all, the majority of these post-movement expressions has taken on the form of witnesses, or oral histories that lack reflectivity, relatively speaking. However, with the tacit understanding between protestors and government that the occupation would not succeed, the ending of the movement by eviction of the protestors from occupied zones was marked as an item of unfinished business, leaving a strong sense of a lack of fulfilment among Hongkongers. The various and profuse post-movement expressions appear as actions of counter-amnesia to keep the spirit of the movement alive.

Art/movement as a public platform

These analyses assert that the Sunflower and Umbrella Movements have contributed to the creativity of the artist-as-citizen and citizen-as-artist to global/local resistance. With the geopolitical 'China factor', the protestors in these two events shared similar sentiments and understandings, and were emotionally connected in their support of each other.

The unfolding of the movements could be seen as a war of visibility. In the Sunflower Movement, the sign of the sunflower was a coincidence, and it carried no deeper meaning and connections with the movement itself. Conversely, in the case of Hong Kong, the motif of the umbrella made the movement distinctively visible by taking on endless manifestations. As to the spatial arrangement, the parliament with its surrounding streets was the main site in the case of Taiwan, while the situation in Hong Kong was more complicated through having several sites with distinctive characteristics. In both cases, the occupations changed the ordinary feature of the sites and transformed them into a living, affective and perceptive place with a memory. Based on the differences of the sites, the parliament could be seen as a temporary gallery producing and presenting various *in-situ* artworks. A large number of the artists' personal involvement differentiate this case from previous examples of resistance. The debates on the art activism and art in Taiwan unfolded in 2014 and 2015. Compared with the case in Taiwan, in Hong Kong there was no heated debate as such. Visualizing political appeals in Hong Kong was via demonstrators' collective making.

After the occupation of parliament by the Sunflower Movement, efforts have been made to change the real political situation in Taiwan. As a result, the KMT with its pro-China stand faced serious failure in the elections, and the DPP with its vision of an independent Taiwan has now been in power since May 2016. For the 'unfinished' Umbrella Movement, the profuse, and at times verbose, post-movement expressions presented in the forms of archiving, publications and art exhibitions, work as mnemonics that bring forth questions about how this movement should be remembered and (re)interpreted in the future.

In these two cases, a new perspective of 'Art/Movement', combined as a whole, could be concluded as a public platform. On this platform, the possibilities of being together can be practiced, collective memory can be presented and affective politics can be experienced. Therefore, the argument of this chapter is that Art/

Movement is a kind of expansion of art, a new form of public art, made by citizen-as-artists and artist – as citizens. This public platform can and would contribute to the imaginations of the society we are in.

Notes

1   With a comparative approach, this paper takes on ideas from the individual papers presented by Lu and Wong at 'Made in Public: Property, Commons and the Alternative Economics of Art' conference, 18–19 September 2015, TCAC, Taipei, Taiwan; Pei-yi Lu, 'Art/Movement as a public platform?', unpublished; Phoebe Wong, 'Exhibiting resistance: Objects in this mirror are _____ than they appear', unpublished.

2   In 2002, the Hong Kong-based research and curatorial collective Community Museum Project (CMP) mounted the exhibition 'Objects of Demonstration: An Exhibition about the Freedom of Indigenous Cultural and Political Expressions' to examine the materiality, visuality and indigenous creativity of protest art. The exhibition later traveled to the Centre for Chinese Contemporary Art in Manchester, UK, in 2004. In an interview in 2016, the CMP addressed the protest objects as 'performative objects'. They are 'performative' because: 'people are able to create and display these objects in processions, deploying ad hoc situational tactics and transforming makeshift materials (perhaps with limited skills) into expressive resources' (Lai 2016: 99).

3   The art critic Oscar Ho used the term 'Creative Expression' (藝述) to substitute the more familiar 'Art/Artwork' (藝術) to refer to the artistic productions made for the Umbrella Movement. This is a Cantonese phonetic twist on the word 'art' so as to shift the emphasis from artistic skills to an eagerness to express. In this chapter, the term 'Creative Expression' is borrowed to refer to both artistic creation (by artist-as-citizen) and collective making (by citizen-as-artist). The meaning of Artist-as-Citizen could be defined as a professional artist who also plays the role of a citizen in a social movement; while Citizen-as-Artist refers to a citizen who creates art in a social movement, similar to a professional artist.

4   The Wild Lily Student Movement held a six-day sit-in at Memorial Hall Plaza in 1990. The demonstrators sought direct elections of Taiwan's president and vice president and new popular elections for all representatives in the National Assembly. The Wild Lily Student Movement marked a crucial turning point in Taiwan's transition to a pluralistic democracy and has since been seen as a model student movement.

5   After the KMT regained power, more social movements emerged which were initiated by a young generation, such as the Wild Strawberry Movement (2008), the Dapu incident (2010), Anti-Nuclear Power (2012-), the Wenlin Yuan urban renewal controversy, the Anti-Media Monopoly Movement (2012), the Death of Hung Chung-chiu incident (2013), when 250,000 people took to the streets to protest, and so on.

6   The OCLP led by the trio Dr. Benny Tai, Chan Kin-man and Rev. Chu Yiu-ming was launched in 2013 as a non-violent civil disobedience campaign demanding democratic elections for the Chief Executive of the HKSAR in 2017. The OCLP believed that the seizing or blockading of the financial and Central Business District – the Central – located at the heart of Hong Kong Island could crack a space for political negotiations with the SAR government, or the Beijing authorities. See www.oclp.hk/.

7   Reflexively, photography critic Lee Wing-ki uses the term 'participatory propaganda' (Lee 2015) to describe the abundant production of derivative works online for protests and social movements, highlighting the very possible nature of these expressions as equal counterparts of authority's propaganda.

8   The term 'affective turn' refers to a new trend in the humanities and social sciences that emphasizes bodily experiences.

9   The artist Chen Ching-yua was a crucial member who initiated and organized people to realize the MV of 'Island Sunrise'.

10 This exhibition was first shown in Japan in March when the Sunflower Movement happened. Artists Chen and Yuan were included. Subsequently, in May, the exhibition toured to the KdMoFA, Taipei, in which these two pieces of work were featured.

11 Over the years, Hong Kong protestors have learnt to equip themselves with masks, raincoats and umbrellas to protect themselves in demonstrations from pepper sprays, which are often used by the police force following the change in the political climate of the city.

12 The 'Lion Rock spirit' originated from a television series *Under the Lion Rock*, produced by Radio Television Hong Kong, beginning in 1972. It embodies 'the belief in solidarity for a better Hong Kong in terms of its economic status, and of equal opportunities for hard-working people to achieve socio-economic advancement'. In her analysis, artist Clara Cheung argues that this banner has redressed the Hong Kong dream: the earlier version of 'Lion Rock spirit' was economy driven and upheld the 'Central Value' (to borrow a term from Taiwanese essayist Long Ying-tai), and was deliberately kept apolitical; the 'Lion Rock spirit' has reinstated a newly found political stand, and 'Hongkongers stay on the Kowloon side to recognize a new battleground' (Cheung, 2016).

13 See www.thelibrarybysoundpocket.org.hk/listen/the-umbrella-movement-field-record-ing-investigative-project/.

References

Augé, M. (1995) *Non-Places: Introduction to an Anthropology of Supermodernity* (J. Howe, trans.), London and New York: Verso.

Chang, S. L. (2014) 'On the visual politics of the Sunflower Movement', *VOP* 13 (September): 24–33 (in Chinese).

Chang, H. H. (2014) 'This is not a sunflower, this is an art action that makes the art down', *ARTalks*. Retrieved from talks.taishinart.org.tw/juries/chh/2014041101 (in Chinese).

Chang H. H. (2015a) 'This is not a sunflower, how the Hill of Chairs could be art?', *ARTalks*. Retrieved from talks.taishinart.org.tw/juries/chh/2015022503 (in Chinese).

Chang, H. H. (2015b) 'This is not art: The reason to nominate Sunflower Movement', *ARTalks*. Retrieved from talks.taishinart.org.tw/juries/hh/2015013102 (in Chinese).

Chen, W. C. (2014) 'Who came to the student movement? The basic pattern of participants in the Sunflower Student Movement', *Street Corner of Sociology*. Retrieved from twstreetcorner.org/2014/06/30/chenwanchi-2/ (in Chinese).

Cheng, J. Y. S. (2014) 'The emergence of radical politics in Hong Kong: Causes and impact'. *The China Review* 14 (1): 199–232.

Cheung, C. (2016) 'Reconstructing the Hong Kong cultural identity by reconnecting with history through art exhibitions and performative rituals (from the construction of the "Lo Ting" myth in 1997 to the revival of ritualistic practices in 2014)'. Retrieved from www.aicahk.org/chi/reviews.asp#sthash.5xzfPquA.dpuf (accessed 23 March 2016).

Erni, J. N. (2001) 'Like a postcolonial culture: Hong Kong re-imagined'. *Cultural Studies* 15 (3–4): 389–418.

Gong, J. J. (2015) 'Sensibility redeployment • haptic visuality'. *ARCO* 272 (May): 104–107 (in Chinese).

Graeber, D. (2014) *Democracy Project: A History, a Crisis, a Movement* (S. J. Tang, S. Y. Li and Y. X. Chen trans.). Taipei: Business Weekly Publications (in Chinese).

Groys, B. (2014) 'On art activism', *e-flux journal*, 56 (June).

Ho, O. (2015) 'Creativity around town, blossom everywhere'. In *Art as Social Interaction: Hong Kong/Taiwan Exchange* (Exhibition catalogue): 92–99. Retrieved from issuu. com/artassocialinteraction/docs/asi (in Chinese).

Lai, S. (2016) 'A conversation with Siu King-Chung about the Community Museum Project', *Yishu* 15 (2): 97–107.

Lau, S. L., Kwong, W. S. and Siu K. Y. (2015) *Love and Justice*, Hong Kong: Up Publications Ltd. (in Chinese).

Law, S. S. M. (2016) *Regarding Umbrella: Creation, Emotion and Memory*, Hong Kong: Department of Visual Studies, Lingnan University. Retrieved from commons.ln.edu. hk/vs_faculty_work/5/ (in Chinese).

Lee, K. (2015) 'Xi Jinping at the "occupy" sites: Derivative works and participatory propaganda from Hong Kong's Umbrella Movement (2014)'. Retrieved from hdl.handle. net/2027/spo.7977573.0006.107 (accessed 15 July 2016).

Lin, F. F. (2014) 'Today's Hong Kong, Today's Taiwan'. Retrieved from foreignpolicy. com/2014/10/01/todays-hong-kong-todays-taiwan/ (in Chinese).

Ma, N. (2015) 'The rise of "Anti-China" in Hong Kong and the 2012 Legislative Council Elections', *The China Review* 15 (1): 39–66.

Minato, C. (2015) *Methods of Revolution: 318 Sunflower as Creative Citizen Movement*, Taipei: Garden Publishing Company (in Chinese).

Solnit, R. (2014) 'The butterfly and the boiling point: Reflections on the Arab Spring and after', *The Encyclopaedia of Trouble and Spaciousness*, San Antonio, TX: Trinity University Press: 22–31.

Thinking Taiwan Forum. (2015) *This Is Not Sunflower Student Movement: The Recording of 318 Movement*, Taipei, Taiwan: Asia Culture Publishing (in Chinese).

Wu, D. K. (ed.) (2014) *Asia Anarchy Alliance*, Taipei, Taiwan: Project Fulfill Art International Co., Ltd.

Weibel, P. (2015) 'Preface', in P. Weibel, (ed.), *Global Activism: Art and Conflict in the 21st Century*, Karlsruhe: ZKM: 23–28.

Weibel, P. (2015) 'People, politics and power'. In P. Weibel (ed.), *Global Activism: Art and Conflict in the 21st Century*, Karlsruhe, Germany: ZKM: 29–61.

Wong, Sampson (2016) Seminar at 'M+ Matters: Confronting activist art and design from a museological perspective' [Presentation], 21 March 2016, Hong Kong.

Wong, Y. D. (2014) 'This is not contemporary art. This is the art harvest that makes contemporary down', *ARTalks*. Retrieved from talks.taishinart.org.tw/event/talks/2014041201 (in Chinese).

Wu, R. R. and Lin, H. H. (eds.) (2016) *Sunrise: The Amplitude, Depth and Horizon of Sunflowers Movement*, Taipei, Taiwan: Rive Gauche Publishing House (in Chinese).

4 Ghana ThinkTank

Mobility, reversal and cultural difference

Gretchen Coombs

In her keynote address at the 2014 Creative Time Summit, Saskia Sassen discussed notions of citizenship, the geography of privilege and the geography of rights. She called for transversality and emergent new architectures for citizenship. Sassen's call to rethink the hierarchical forms of thought in relation to nationalism, migration and the stigma surrounding immigration emerge through a more pronounced transversality, that is, actions and language that would cut across these vertical structures (Sassen 2014). Following her keynote, several artists and activists described how they embrace Sassen's challenge. They were contemporary artists and collectives who confronted a range of social and political issues through short-term interventions in public spaces or long-term engagements with targeted communities. They presented strategies to address issues such as immigration, labour rights, food access and incarceration, that might include performance, education, activism and acts of reversal. Their tactics to achieve these involved tea ceremonies, mapping contested histories, tactical magic and readymade vehicles. One of these groups was the American-based collective, Ghana ThinkTank (GTT).

GTT is comprised of a group of four artists (Christopher Robbins, John Ewing, Matey Odonkor, Carmen Montoya) that have been 'Developing the First World' since 2006. Their socially engaged, participatory projects involve 'collecting' problems from selected communities in the United States and Europe, which are then sent to their global network of 'think tanks'. These think tanks take various forms, including, for example, a group of bicycle mechanics in Ghana, a radical radio station in El Salvador, artists in Iran, groups that have multiple degrees of separation from urban centres in the United States. As problem solvers, GTT flips the conversation on international development, not to undermine the value of non-governmental organisations' (NGOs') work per se, but instead to challenge the very assumptions that perpetuate the 'First World' and 'Third World' construction.

During the Cold War, and with the emergence of development/international aid, the binary of 'first' and 'third' world gained traction. Countries of the global south (third/developing world) were incorporated into a global political economy, dominated by the global north (first/developed world). Development interventions were designed to bring these 'developing' nation states into the modern world. Even though development is now a set of contested values, applications and outcomes, its legacy comes from colonial and Orientalist assumptions of the other (Said 1987; Bhabha 1990; Hall 1997; Hodder 2007), which constructed the narrative of 'civilised/primitive'. Therefore, in the context of development, this binary worked in tandem with the hero/helpless narrative that has underpinned much of the rationale for international aid (Robertson and Hite 2000).

Through artistic processes that ultimately reveal cultural differences and shared meanings, GTT's projects highlight how an artist collective's social practice becomes situated in local, international political, economic and social contexts, and as a result, challenge the common assumptions about development.[1]

According to the group's website:

> These think tanks analyze the problems and propose solutions, which we put into action back in the community where the problems originated – whether those solutions seem impractical or brilliant. Some of these actions have produced workable solutions, but others have created intensely awkward situations, as we play out different cultures' assumptions about each other.
>
> (GTT 2015)

In one of the most oft cited quotes in the field of socially engaged art, Tania Bruguera has famously described what she does as an artist, 'I don't like art that points at the thing. I like art that is the thing' (as cited in Dolnick 2011). Arguably, GTT does something similar: they become the thing, a 'new type' of development, while simultaneously pointing to the thing, the fallacy of development. And if GTT's ultimate strategy is to 'Develop the First World', then they use the tactics of reversal and mobility to achieve their goal. Their work is threefold: they use a modified readymade vehicle (a tear drop trailer, donkey cart) that acts as a collection point for the problems; run workshops and think tanks that find solutions and; create gallery installations, often conceptual, that reflect process of collecting and solving problems. The art gallery installations maintain the register as art. When the problems are collected and then sent to think tank for solutions, and then the solutions implemented back to where the problems originated, the development discourse begins to shift.

Collective member Christopher Robbins stated: 'Part of the agenda is to point at the unintended consequences that outsider solutions can create, while another part is to demonstrate that the rest of the world has something to offer' (Robbins and Kennedy 2012). GTT's projects usually start with a commission from an arts institution. From there, the group spreads out into the surrounding area to look for stereotypes that stem from cultural, social or political issues circulating in that community. They have worked in towns and cities such as Westport, Connecticut; Detroit, Michigan; or the Serb and Albanian populations divided by the river in Mitrovica, Kosovo. Members then devise a way to collect problems from that area and deliver them to think tanks, which are generally at a geographical distance. Participants engage in a 'problem-solution' process and encounter multiple lived worlds. The results leverage the visibility of subaltern groups, shift the media ecology of representation, and create cross-platform dialogues that ultimately reveal commonalities. They move between and across real and imaginary boundaries to cut through the established systems that maintain power dynamics. These interventions subvert the hero/helpless narrative that supplanted the civilized/primitive construction during the process of decolonization.

Although the collective has numerous projects from their nearly decade of work, this chapter focuses on two projects – the Mexico Border Project and the Corona/Queens Project. Their tactics of reversal and mobility in these projects address immigration, the controversial and complex system that has become a moving target on the American political landscape, in which the looming threat of a border wall dominates political theatre. Neither reversal nor mobile structures are new in art practices, yet it is worth reviewing a few notable projects that have affinities with these GTT projects.

Reversal

Art practices have offered some of the most potent interventions in the legacy of anthropology and colonialism which located, identified, classified, and constructed a 'primitive' other 'over there' (Fabian 1983; Torgovnick 1990; Hall 1997). These racialised assumptions are perpetuated through institutional processes like international development as well as in the media. Guillermo Gómez-Peña, the Mexican-born performance artist has, since the 1980s, been engaged in 'reverse anthropology', staging 'postcolonial' performances (along with his performance troupe La Pocha Nostra) that foreground race and intervene in cultural fears and desires by focusing on obsessions with the exotic. He uses elaborate performative and interactive elements as he did with then collaborator, Coco Fusco, when *The Couple in the Cage* (1992–1993) performance travelled worldwide to expose (to the audience) their deeply embedded cultural stereotypes and desires for the other. More recent performance work such as *Ex-Centris* (2004) and *Corpo/Illicito* (2011) where Gomez-Peña and members of La Pocha Nostra used their bodies to deliberately act out their cultural otherness (shaman, warriors, prostitutes) so that the audience would encounter and reflect on their own desires.

Arguably, GTT employs similar tactics, albeit not undermining at the level of racialised desires. Their work operates on the functional end that maintains the other's problems – poverty, lack of potable water, access to healthcare, for example – be rectified through aid interventions. Their tactics draw attention without explicit acknowledgment that the cause of some of these very real problems is the fallout from centuries of colonial rule. The group does not suggest that all development is bad, but wants to challenge the assumptions at its core. As Robbins put it: 'We all saw the other side of good intentions' (Banff 2016). With GTT, this assumed power dynamic gets flipped and local knowledge is valued by foregrounding context – the 'hero' gets 'help' in the form of a solution to their (self-defined) problem. Several of the group's online videos reveal one potential critique of GTT's projects; that is, the assumption – and possible imposition – that those in the 'developing world' should be burdened with first-world problems in the first place. In his video interview with Sue Bell Yank for Banff Centre for Arts and Creativity, Robbins described how some of GTT's critics claim that the choice to use think tanks is either condescending or presumptuous, and further highlights the vast inequality and uneven power dynamics. In response to the assumption about the assumption, a video of Indonesians offers a rupture to this narrative: think tank members indicated they have experience with problems so maybe they are the right people to respond (Banff 2015). Robbins, a former Peace Corps volunteer in Benin, Africa, acknowledges how his attempt to reverse development implicates him in a different process of development, but hopes his work can begin to unravel cultural stereotypes.

Art and the age of mobility

Since the notion of mobility characterises the modern age it is no wonder that artists have also used various types of vehicles in their work to disseminate information, reach beyond urban hubs, to suburbs, community centres and other sites not otherwise visible. Some of the more notable works in the last decade have included Studio *REV-'s NannyVan* (2014–2015), part mobile design unit, part recording studio which travelled around US cities stopping at museums, libraries or other sites frequented by domestic workers and their employers to offer domestic labour rights information. Artist, Jon Rubin chose the residential neighbourhood of Hillman City, Seattle for the *FREEmobile* (2003), a modified 1976 Chevy van, which moved through the neighbourhood so that residents could exchange products and services with each other, familiar and unfamiliar. More politically charged, the Center for Tactical Magic's *Tactical Ice Cream Unit* (2008) used a large white van reminiscent of a communist-era spy vehicle that had a dual purpose; the truck supported protests with a legion of surveillance cameras that could monitor police activity, as well as offered 'treats for the streets', ice cream with a side of propaganda. And William Pope L.'s seminal work, *The Black Factory* (2005), brought the issue of race – through staged artefacts and pedagogical interventions – to street corners and museums across the country.

Like those mentioned above, GTT has used mobile structures to move through different areas. GTT's ready-mades (manufactured objects that an artist selects then modifies) traverse visible and invisible borders to remote towns or villages often associated with the developing world or represent contentious geopolitical sites like the US border. The collective is invested in the way their materials (vehicles in this case) become tools for the artistic process. Moving through a neighbourhood, between several communities of participants, improves communication and increases the complexity of each project and responds to local context, not always a consideration in international art projects.

Instead of being seen in an art museum context only, the structures moving along the streets made the project more accessible to a broader swath of people. Collective member Carmen Montoya suggested: "It shifts the language in which the project can be interpreted – a mobile workshop or a coffee/tea dispensary, a community watering hole." In Morocco, the donkey cart also functioned as a tea lounge, with pillows and a video camera for people to give their statements, accommodating culturally specific customs such as gathering for tea. Its role as an art object fades into the background. Coupled with the moving structures they make that are collection points for problems, GTT formed the think tanks for each project, which are usually remote and distant from the point of collection. The groups of people that formed the think tanks provided the solutions using local knowledge, experience, and their interpretations of context in which those problems are generated (for example, the Westport, CT project). For GTT, this distance makes cultural difference more apparent.

Immigration two ways

Immigration is a key theme in socially engaged art, and GTT's projects in Corona/ Queens and the Mexican border work in tandem with other artists working in this area. Artists such as Tania Bruguera and her Immigrant Movement International (IMI) (2011-) project confronted the complex social issues surrounding immigration. In Corona, Queens (New York), she created a centre for immigrants and their families to gather, get legal help and take English classes. This long-term project, funded in part by Creative Time, a New York non-profit arts organisation, and done in conjunction with the Queens Museum, has become a gathering place and an educational centre. Critiques of the project included Ellen Feiss, who outlined the oversimplification of 'rights' as they are disseminated at IMI, and Andrew Friedman, director of Make Road, New York, who raised suspicions of an artist who can exit a context at will (as cited in Dolnick 2011).

A more symbolic gesture is artist David Smith's flight over the border between Tijuana and San Diego. Smith was shot out like a human cannonball – launched himself through the air – a metaphorical trace flying over the border. He flew above the law, and subverted the order of the border control so heavily monitored on the US/Mexico border. The migrant youth group, La Ruedada, gathers stories of migration and brings them to public spaces – often through a choreographed dance that the group interpreted from an experiential activity about their migration – to help build community knowledge across generations and ethnicities.

These varied projects reveal some of the debates circulating about socially engaged art: the short-term versus the long-term nature of a project, the ethical versus the aesthetic, and more pertinently for GTT, 'is it useful?' (Thompson 2012, p. 16).[2] This distinction intersects with Bruguera's quote and how GTT uses their form to confront immigration at this critical moment in US political history. Being a 'thing' and pointing to a 'thing' collapse when we ask, 'is it useful?' How it is useful depends on if you ask a participant, a member of the think tank, someone from the artworld, or one who works in development.

Corona, Queens Project

Roosevelt boulevard stretches across Queens, New York, and traverses through one of the most ethnically and culturally diverse areas in the nation. Several waves of immigrants from Europe, Latin America and Asia have all made their homes around this central hub in Corona, Queens, a neighbourhood surrounding Flushing

Meadows Park (site of the Queens Museum and two World Fairs). At first glance this might represent the all-American melting pot so carefully crafted as part of the American imaginary, but a closer look reveals what often is ignored in representations of diversity or even immigrant communities; that is, with each wave of immigration and subsequent assimilation, comes the very real tensions that exist between the 'older' migrants and the newer, less established groups. The racial and social tensions through waves of immigrants/assimilation processes became the focal point for GTT's work in Corona, Queens.

Commissioned in part through Creative Time and the Queens Museum, GTT used a custom-built drop trailer that for three months in 2011 travelled into different communities in Corona. The project took place over three months in 2011 during which time problems were collected inside the trailer and sent to think tanks in Gaza, Iran, Lebanon, and Serbia. The trailer was an important way for the Queens Museum to extend their work into the Corona community. According to Montoya, many people are undocumented and not a museum-going audience, so 'we took the trailer to meet them where they did their daily gatherings, park plazas, on the streets – it seemed not only natural but necessary'. As a mobile workstation, the teardrop trailer included a bench, carpet, papers, two or three video screens and paperwork. The back of the trailer featured a screen with a large red button to record your problem. The entire area we focused on within Corona, Queens, was contained in a rough circle approximately one mile in diameter. The group chose sites for their likelihood of having people ready to engage with the cart (e.g. a shady spot under a subway station where people wait), or popular sites like The Lemon Ice King of Corona (where Italian ice cream has been made for sixty years) and 'Spaghetti Hill' park (see Figure 4.1).

Figure 4.1 The Ghana ThinkTank Mobile Unit in 'Spaghetti Hill' in Corona, Queens. Copyright Ghana ThinkTank.

They also chose sites to best work with their partners of the overall project and enactment of solutions such as with Queens Museum's Passport Fridays (an entertainment event) and Immigrant Movement International events. Each stop offered different views of race and social relationships, and made visible power tensions between different waves of immigration. Problems described included: 'I feel like a minority in my own neighbourhood' and 'The old-timers aren't welcoming to newcomers'. More specifically, many of the more established Italian immigrants took issue with the massive influx of Latino immigrants. And many of the Ecuadorian immigrants took issue with the influx of Mexican immigrants. These tensions suggested there were internal conflicts within the diverse immigrant communities.

Montoya described how this part of Queens is a fragile site; that is, there are many languages spoken, political affiliations, immigration statuses, and so on, so, the initial outcomes were not what GTT intended. Together they asked, 'how can we make more nuanced, granular work?' The group then made the decision to have the whole project in Queens since this nuance couldn't be solved anywhere else, which was something the group sensed they overlooked when forming the think tanks in geographically distant sites. They realized that the 'flip' of the power dynamic was not working in Corona, so 'Developing the First World' no longer worked in the context of this diversity. They spread the power dynamic locally, with newer Latin American/Caribbean immigrants solving the problems of established Italian immigrants during a series of workshops at Bruguera's Immigrant Movement International.

The workshops and focus groups allowed them to bridge the immigration experience between generations of immigrants (see Figure 4.2). One resulting solution included asking people who expressed anti-immigrant rhetoric to design

Figure 4.2 Recent immigrants from Latin America solving problems of third generation immigrants from Italy in Corona, Queens, at Tania Bruguera's Immigrant Movement International. Copyright Ghana ThinkTank.

pro-immigrant ads to be displayed in their own neighbourhoods. Made to resemble a proper bus ad, one ad placed on the back of the bus featured a man with the quote 'I came here to be American' as it travelled on the Q23 bus line through the striated space of Corona. Another showed the Brooklyn Bridge over the statement 'Made by Immigrants'. This slogan came from the grandchildren of Italian immigrants talking about how their grandparents had 'built this city!' Montoya describes another solution from these workshops, Legal Waiting Zones:

> Ticketing pedestrians for "loitering" has been illegal in New York state since 1983. The Black and Latino residents of Corona, Queens who are the victims of racial profiling by police don't know that. This "Legal Waiting Zone" sign was based on her interviews with people on Roosevelt Ave. and is an effort to help raise consciousness about the issue.
>
> (Montoya 2015)

One sign says 'legal waiting zone: it is ok for you to wait here, and in all public places, for a friend, your mom, or simply because it's too hot in your apartment'. Another solution/action, based on a solution from the think tank of incarcerated girls, and modelled on performance artist Adrian Piper's *Calling Cards* (1986–1990), involved distributing greeting cards to people who were often racially-profiled by police.[3]

The Corona project shifted from a critique of presumed power dynamic based on divergent geographical features into a more nuanced understanding of immigration in the United States, where tensions between communities are often missing from discussions on immigration. These invisible borders give way to homogeneity and diversity dissolves into sound bites for political gain. But the project only lasted three months, akin to an art world exhibition cycle. The enacted solutions remain as art objects.

The Mexican border

As part of a three-year project (2013–2016), GTT's work at the Mexican border approaches the lightning rod of contemporary politics, and the divisive nature of immigration. Because of this oppositional political landscape, cultural assumptions prevail – 'illegals', 'job stealers' – and more racialised terms uphold more deeply entrenched tensions on the San Diego/Tijuana Border, which forms the largest binational conurbation on the US/Mexico border. This extended urban area resulted from several towns merging with the suburbs of Tijuana and San Diego. Part of this merge comes from increased transportation nodes that linked the areas to form a single labour market, and often function as part of the same city where school children and workers cross the border daily, specifically at the San Ysidro Point of Entry. In some respects, this area has become ground zero for many of the debates about immigration in the United States, and understanding this area's congestion – over 12 million cross a year, 300,000 daily commuters often with four to five hours of waiting – gives migration flow a different meaning. The binational art collective, Cognate Collective, occupied this plodding border crossing with their 2014 work, *Dialogue in Transit-Evolution of a Line,* where a caravan of vehicles staged a 'conference' where artists and activists reflected on the North Atlantic Free Trade Agreement (NAFTA) and imagined new forms of border crossings.

Because of the contentious nature of the region, and because the groups are not so geographically distinct, Robbins describes how in this project (based on one in Kosovo) they create 'enemy think tanks', where there are groups in conflict – those crossing the border, recently deported immigrants, the 'Minutemen' (a civilian border patrol from the Minutement Project) (Banff 2016) – asking think tanks of recently deported immigrants in Tijuana and undocumented (migrant) workers in San Diego to solve the problems of Minutemen, patriot and nativist groups. He describes how one Minuteman's problem was that immigrants were taking his jobs, undercutting his wages. Another complained that he felt like a minority in his own neighbour-hood, and needed to learn Spanish to find a job in his own, English-speaking coun-try. The think tanks (comprised of recently deported immigrants) responded that the two groups (migrants and Minutemen) weren't against each other, but were actually being pitted against each other for the lowest wage. So, really the problem was the owner of the business who wanted the cheapest labour. One solution outlined the need for a union of legal and illegal workers to demand fair wages for everyone. GTT worked with think tanks for more specifics so the Minutemen could implement them in real and substantiative ways, such as working with the immigrants, rather than allow their employers to pit them against each other to drive down wages. However, GTT members related how during the 2016 presidential campaign, it was more difficult to work with the Minutemen who had become comfortable being outwardly racist, citing Donald Trump's anti-immigrant stance.

Many who constitute the border populations are in precarious situations and have a vague relationship with the law. Robbins relates how often think tanks wanted to remain anonymous and met in private homes in Tijuana to discuss solutions. GTT searched for a more public site to engage with people about the border and realized 'a painfully obvious place: the border!' The think tanks were recently deported immigrants that helped solve issues of mobility. The group real-ized that when they started working on issues related to the border, that mobility (or lack of) was built into the flow and crossing between Tijuana and San Diego. The pedestrian line can take several hours to cross. During workshops GTT used Participatory Action Research (PAR)[4] techniques in collaboration with the Tijuana-based art collective Torolab and several think tanks in Mexico to develop a brightly-coloured border cart with green and blue wheels and red and purple pivoting seats under a shade structure.

Attached to each seat was an iPad, an invitation to interact. GTT used this cart as a collection point for people to submit problems about immigration (if they are 'American') or solve those problems (if they are 'immigrants'). The cart was equipped with a modified winch controlling the spikey wheels that slowly advance people along the pedestrian line at the same slow pace. And while people sit and cross the border, they deal with contentious issues directly across from each other. Everyone in line has the same goal – moving across that border – but is separated by many other aspects of that border they are crossing together.

Robbins describes how the design and colour was based on 'Tablitas', a folding wooden block toy, which is similar to a Jacob's Ladder toy (see Figure 4.3). The cart is designed to fold and stack up flat like those toys, and then pop open, which allowed GTT to deploy it to other cities dealing with divisions between groups of people, such as the 'gentrified' and 'gentrifiers' of East Austin (Texas). However, there was a double register of mobility at the border – one that uses mobility, a slow moving cart, to underscore the difficulty in crossing the border in an easy and efficient way. At the same time, the cart offers those in line an opportunity to interact, and moving through the line and across the border gives a depth to the project that wouldn't have happened if it had remained stationary. Robbins states:

> We have realized we fit in more at the border if we charge, so for a small fee people crossing the border can sit down, watch movies, and solve (or submit) problems, as we push the wooden cart across the pedestrian line. I am particularly looking forward to the moment when I tell the border guard that we don't know each other – he just paid me a few dollars to push him across the border, which makes me an illegal worker (2010).

The longer-term nature of the Tijuana Border project brings a different set of expectations and methods. No longer are they tied to the short cycles that organize art gallery and museum commissions and subsequent exhibitions. The group becomes more accountable to their mission and goes through the process of slowly building alliances in the areas in which they work with other artists, NGOs and advocacy groups who share similar goals. They become less like an art collective and more like an organization, with administrative roles and the like. As Robbins

Figure 4.3 The border cart in Tijuana, a hand-cranked vehicle with facing seats and custom software designed to create public think tank sessions at the US/Mexican border.

states about the Border project: 'We fold back into what it was we were questioning.' They hope to figure out an exit plan by making room for people there to take over (Banff 2016). The stakes are higher when there's a longer-term project, more relationships, more structure, more risks to the fragility of the lives they touch. While they might become the 'thing' they wanted to subvert initially, their art installation can continue to point to 'the thing'.

Conclusion

Returning to Thompson and Bruguera (2011, para. 23), art being a thing and pointing to a thing, and asking about its usefulness presents a complicated conundrum for socially engaged artists. No doubt any critique levelled at GTT's work requires some careful unpacking, as any challenge to the fixed conceptions and assumptions about the other and the world are fraught with uneven and uncertain outcomes.

Development carries decades of policy and implementation, and the symbolic and real interventions the group tackles can't be easily understood out of context. 'Developing the first world sounds witty in Westport, Connecticut but in Detroit, Michigan not so much, it sounds true' (Banff 2016). Robbins' comment encapsulates some of the potential critiques of GTT's work – understanding context is crucial, and there are areas not usually associated with the 'developing' world that carry the same stigma. The same things that inspired the group are some of the same things that can be critiqued in their work – that uneven power dynamics remain, they are artists and choose to enter and exit, they may challenge cultural bias, but create new ones. 'Developing the first world' might be a subversive practice in that it challenges the continued neocolonial notion of progress in a globalised world. It may also have hidden implications that further perpetuate how we might understand this reframing. In other words, is a subversion of the developed/developing world enough? Or, does it become only symbolic, with its efficacy questionable beyond the discursive sites of the art world (Amin 2013).

Current work like the Mexico Border Project (2016), in which the group partnered with NGOs and development agencies, will open up the possibility that some of their work can be applied in a more practical manner. NGOs and other industries, such as banks invested in improving the lives of immigrants or other subaltern groups, might appropriate what is useful to them from GTT's methodology. It would be interesting to see how work such as that on the Mexican Border can be seen in the context of 'service design', where the concepts created present alternatives to existing service structures, specifically in the public sector.

What is at stake in their work has grown over time, and if the group's goal is to become relevant in other contexts, the Mexico Border Project seems to be moving in that direction. If so, will this then become prescriptive? Can their work be at once critical of development but also be used as a 'new form' of development without shifting the structural and representational values that underpin the very notion of development itself? Even contemplating this knowledge transfer and subversive methodological approach can over burden the artistic process that gives much life to GTT, and risks instrumentalising the process so much so that its strains its 'value' as art. This is where we can go back to pointing at the thing: if their work were actually incorporated through NGO work, then scale would matter. And then it would become the unwieldy structure that it already is.

GTT's aesthetic and critical interventions involve collecting problems and deploying solutions. This exchange becomes a 'new' form of engagement for groups who 'deposit' their problems, the think tanks who devise the solutions, and again for those who have to enact the solutions (those who originally 'deposited' the problems). Such a process might begin to tackle some of the biggest assumptions that different cultural groups have about one another (challenging individual versus the collective; the notion of what rights means). So, ultimately, the group doesn't solve problems, they never claim to. The artistic efforts through reversal and mobility of socially engaged artists like GTT can continue to operate in a shifting framework of perception and application. Their material and conceptual process might just allow an erosion of cultural bias through these types of encounters, an outcome that Guillermo Gomez-Peña describes as a '3rd Border Zone', an experience where participants realize borders are imaginary. It is a realisation that opens up endless possibilities for inclusion, empathy and promoting new intimacies between divergent cultural groups.

Notes

1 Unless otherwise noted, all quotes from Robbins and Montoya are through personal correspondence (November 2014 and July/August 2016).
2 Nato Thompson's reconfigures question of efficacy in socially engaged art projects changing the 'Is it art?' question to 'is it useful?'
3 The cards, meant to be handed to police officers by those being racially-profiled, read:

> *Esteemed Officer, I understand that it is your right to stop and question me. Clearly you feel that I have given you some reason to doubt the integrity of my behavior. After all, I have been standing here for some time with no apparent reason for doing so. Then again you know, as I know, that there is nothing necessarily criminal about this—but perhaps it is my age, race and gender that concern you. I am well aware that over 84% of the allegations of police misconduct made in Corona, Queens involve young black or Latino males.*
> *I fit the profile.*
> *This is foremost in my mind as you approach me. I want you to know that I respect the premise of your role and that I acknowledge the risk of personal peril that you undertake each time you report for duty. But you should also know that your rigid stare evokes in me a sense of vulnerability and apprehension that seems out of place in our relationship: you an officer of the peace and me a member of this community you swore to serve and protect. Regardless, here we are and we must make the best of this unpleasant circumstance.*
> *I assure you that I will behave in a deferential and cooperative manner while noting the details of our encounter: your name, badge number, and patrol car number. I ask only that you treat me with courtesy, professionalism, and respect.*

4 Participatory Action Research is a method which requires research and action 'with' communities, not 'on' or 'for' the community.

References

Amin, Ash (2013) 'Telescopic urbanism and the poor'. *City (London, England)* (1360–4813), 17 (4), 476.

Banff Center for Arts and Creativity (Christopher Robbins in conversation with Sue Bell Yank). Published April 5, 2016. www.youtube.com/watch?v=mCzbKEAZ4I8. Accessed May 31, 2016.

Bhabha, Homi K. (1994) *The Location of Culture*, London: Routledge.

Dolnick, Sam (2011) 'An artist's performance: A Year as a poor immigrant'. *The New York Times*, 6 May 2011. www.nytimes.com/2011/05/19/nyregion/as-art-tania-bruguera-lives-like-a-poor-immigrant.html?_r=0. Accessed June 23, 2016.

Ghana ThinkTank, www.ghanathinktank.org.

Hall, Stuart (1997) *Representation: Cultural Representations and Signifying Practices*, London: Sage Publications.

Feiss, Ellen (2016) 'What is useful? The paradox of rights in Tania Bruguera's "Useful Art"'. *Art and Education*, www.artandeducation.net/paper/what-is-useful-the-paradox-of-rights-in-tania-brugueras-useful-art/. Accessed July 1, 2016.

Montoya, Carmen and Robbins, Christopher. Personal correspondence, 29 November 2014; July, August 2016.

Montoya, Carmen (2015) www.lajunkielovegun.com/MariaDelCarmenMontoya/

Roberts, J. Timmons and Hite, Amy (2000) 'Editors' introduction', in J. Timmons and Amy Hite (eds.), *From Modernization to Globalization: Perspectives on Development and Social Change*, Oxford: Blackwell, 1–23.

Robbins, Christopher and Erin Sickler (interview) (2010) 'Pardon this brief commercial interruption: Ghana ThinkTank', *Art 21 Magazine*, blog.art21.org/2010/06/06/pardon-this-brief-commercial-interruption-ghana-think-tank/#.UuXORmQo53J. Accessed 12 January 2016.

Robbins, Christopher and Channing Kennedy (interview) (2012) 'Kony this: "Ghana ThinkTank" turns the tables on White saviors', *Colorlines*. www.colorlines.com/articles/kony-ghana-thinktank-turns-tables-white-saviors. Accessed 18 July 2016.

Sassen, Saskia (2014) Keynote speech at Creative Time Summit. creativetime.org/summit/2014/11/14/saskia-sassen/. Accessed March 30, 2015.

Thompson, Nato (2012) *Living as Form: Socially Engaged Art from 1991–2011*. New York, NY: Creative Time, 16.

Torgovnick, Marianna (1990) *Gone Primitive: Savage Intellects, Modern Lives*. Chicago: University of Chicago Press.

5 Diversifying the stage of policymaking

A new policy network in Berlin's cultural field

Friederike Landau

Unpacking shifts in Berlin's cultural policy field

The composition and constellation of Berlin's cultural protagonists have drastically changed since the early 2000s, when Berlin's rise as an international hotspot for contemporary art production was coined by Mayor Wowereit's (in)famous slogan 'Poor, but sexy' (*Spiegel Online* 2014). Poignant events such as the abolition of the art fair *Art Forum Berlin* (2011) and the debate about a *Kunsthalle* (art hall) for Berlin have politicized the artistic scenes. Attempts to institutionalize collaboration between the players of Berlin's cultural field have brought about controversial events such as the cultural summit 'K2' (2012), initiated by the *Senatskanzlei für Kulturelle Angelegenheiten* (Senate Chancellery for Cultural Affairs; cultural administration) as well as genre-specific forms of dialogue such as the *Jour Fixe* with various visual arts stakeholders, lobbied for by *Netzwerk freier Berliner Projekträume und initiativen* (Network of Berlin Independent Project Spaces and Initiatives). These meetings led to incremental material gains for the independent scene, such as the introduction of a prize for project spaces in 2013. The independent scene shall in the following be understood as the totality of all freely producing, Berlin-based artists, ensembles, facilities and structures in free sponsorship from the realms of architecture, visual arts, dance, drama, performance, new media, music – ranging from baroque, electro, jazz, classical music to new music – musical theatre, children and youth theatre, literature as well as all other inter- or transdisciplinary forms of cultural production (see Kucher 2013: 7, translation: FL).

The complicated, yet continuous relationships between the cultural administration and the artistic scenes were significantly restructured when the transdisciplinary *Koalition der Freien Szene* (Coalition of the Independent Scene; *Koalition*) entered the cultural political stage in 2012: the *Koalition* bundled existing genre-specific claims and forms of organization, and collectively requested reforms for both existing cultural funding as well as new funding instruments and structures (for a more detailed description of *Koalition*'s organizational profile see Landau 2016a). *Koalition* has a broad base of 'member' associations, even though neither individual nor collective membership has been formally institutionalized, delegating speakers from the visual arts, performing arts, literature, music (jazz and new music) as well as artist-run centres (*Kunstvereine*), transdisciplinary arts and project spaces. *Koalition* has issued a public campaign (see Landau 2015) and has actively taken part in budgetary negotiations.

This chapter highlights the interactions between the cultural administration and the *Koalition* as a transdisciplinary collaborative policymaking process amongst Berlin's cultural stakeholders,[1] and seeks to contribute to the understudied phenomenon of what role artists or artist-led organizations play in cultural policymaking processes or cultural governance (building upon studies such as Woddis 2013).[2] It brings forward an empirically-grounded study of the collaborative policymaking process of 'CityTax15', the first and singular distribution of funds from the City Tax, a 5 per cent levy charged on tourist overnight stays, introduced in January 2014 (Senatsverwaltung für Finanzen 2015a). The distribution of funds had been negotiated between the cultural administration and *Koalition der Freien Szene* throughout 2015. Seeing collaboration as a 'way to establish new networks among the players in the system and increase the distribution of knowledge among these players' (Innes and Booher 2003: 36), the chapter investigates the relational, procedural and institutional entanglements of the policymaking setting of CityTax15. I argue that exchanges of various resources between the artist organization, *Koalition*, and the cultural administration led to the construction of a concrete policy outcome, which created a new policy network. The findings from this exemplary Berlin case study might offer new insights about how changing political constellations between cultural stakeholders affect the meaning and relevance as well as the funding practices and realities of cultural and creative cities today. The emergence of self-organized collective agents like *Koalition der Freien Szene* can unsettle and pose challenges to the conventional institutional and procedural settings of policymaking. Hence, the growing mobilization of cultural actors necessitates both empirical and theoretical investigations for and from scholars of cultural policy studies, cultural sociology and geography, social movement studies and political theorists interested in questions of democratic deliberation and governance. Extending the focus of policy analysis beyond the emergence and articulation of a publicly conceived problems, that is, the agenda setting capacities of civil society actors (Kingdon 2011; Sabatier 2014), towards a more relational and reflexive investigation of emerging forms of collaboration[3] and policy formation between the cultural policy stakeholders, will enhance our understanding of the development of potentially new forms of cultural governance (e.g. Peck and Theodore 2012; Jessop 2006).

Policy network analysis

Due to the interdisciplinarity of the field, I engage with literatures on new social movements (e.g. Haunss 2004; Baumgarten *et al.* 2014), dealing with political opportunity structures (McAdam *et al.* 2001; Tarrow 2011) and framing activities (Benford *et al.* 2014), together with approaches from urban governance (Pierre and Peters 2000; Pierre 2005). To breach the rather state-centric understanding within existing models of urban governance (e.g. DiGaetano and Strom 2003), which do not comprehensively include self-organized actors in the formulation of policy agendas, and at times reproduce binaries such as either adversarial or collaborative policymaking, or even more problematically, governance either as top-down or participatory (e.g. Fung *et al.* 2003: 262). My theoretical framework is complemented by concepts from critical policy studies. Following an interpretive and post-positivist approach (e.g. Hay 2002; Fischer 2003; Hajer and Wagenaar 2003; Fischer and Gottweis 2012), I investigate the policy-relevant groups which 'become "interpretative communities" sharing thought, speech, practice and their meanings' (Yanow 2003: 237). Or, similar to the idea of governing coalitions, which I take to be 'the means by which urban political actors seek to define, shape, and implement policy agendas' (DiGaetano and Strom 2003: 371'), I discuss the newly emerged policymaking entity at hand with a framework of policy network analysis (Rhodes 2007; Marsh and Smith 2000). Ultimately, policy network analysis allows for a meso-level investigation of the relationships between state and society which fits the empirical case of CityTax15.

Policy networks are 'sets of formal and informal institutional linkages between governmental and other actors structured around shared interests in public policymaking and implementation. These institutions are interdependent. Policies emerge from the bargaining between the networks' members' (Rhodes 2007: 1244). Policy networks are both an agent-driven practice as well as institutional(ized) materiality where policy settings are constituted and developed. Policy as reflexive practice 'inscribes itself into its texture, and creates/rewrites order by drawing from a multitude of discursively available narratives, modes of representation, artefacts and technologies' (Gottweis 2003: 261). Thus, the novel emergence of the policy network between the cultural administration and *Koalition der Freien Szene* is conceptualized as a communicative, discursive and material institutional terrain within which CityTax15 was negotiated. This 'flexible and dynamic strategic alliance' (Hay and Richards 2000: 2) is considered not only as arena of policy innovation or effective or 'successful' policy implementation, but also 'analysed and indeed appreciated as site[s] for the articulation of conflict and difference, as a place of social and cultural contestation' (Hajer 2003: 99).

At the heart of policy network analysis, Rhodes stresses the importance of the exchange of resources between actors, culminating in the observation that the 'relative power potential is a product of the resources of each organization, of the rules of the game and of the process of exchange between organizations' (Rhodes 2007: 1245). The exchange of resources is necessary for organizations to achieve their goals. These goals do not have to be identical, but constitute 'complementary needs and assets' (Wohlstetter *et al.* 2005: 421). Furthermore, within policy network analysis, the binary of policy communities and issue networks falls short of tackling the empirical phenomenon at hand (Marsh and Rhodes 1992; Fawcett and Daugjberg 2012). The former is characterized by a limited number of participants who share a common political agenda, but are co-dependent in order to achieve their goals (Fawcett and Daugjberg 2012: 199). In contrast, issue networks are represented by a large number of participants, 'fluctuating interaction and access for the various members; the absence of consensus and the presence of conflict; interaction based on consultation rather than negotiation or bargaining; an unequal power relationship in which many participants may have few resources, little access and no alternative' (Rhodes 2008: 428). Besides the difficulty of determining a small or large number of network members, the limitation of this framework lies in the ontologically narrow differentiation between either consensus or conflict over policy principles. Such a dichotomous stance is incompatible with my post-foundational and agonistic understanding of politics, which assumes that human relations are inherently and irrevocably conflictual, and that consensus can only temporarily be fixed (Mouffe 2005).

Moreover, the construction of a new policy community and resulting policy-making practice, links to the idea of the institutional void, which assumes that 'there are no clear rules and norms according to which politics is to be conducted and policy measures are to be agreed upon' (Hajer 2003: 175). Thus, the encounter of policy actors in an institutional void leads to 'negotiate new institutional rules, develop new norms of appropriate behavior and devise new conceptions of legitimate political intervention' (Hajer 2003: 176). Here, the *Koalition der Freien Szene* in Berlin could be seen as a policy entrepreneur whose 'presence and actions can significantly raise the probability of legislative consideration and approval of policy innovations' (Mintrom 1997: 738), who entered an institutional cultural political void.

As the controversy around the explanatory power of policy networks continues (Dowding 1995; Börzel 1998), I neither ascribe causal power regarding the policy outcome solely to the policy network's actors, nor wholly to the structural characteristics of the network. Instead, I adopt a dialectical approach to policy network analysis (Marsh and Smith 2000). This allows for the investigation of the interpenetrating interactions between policy outcomes, (policy) contexts, the formal and internal structure of the network, and its agents (Marsh and Smith 2000: 5). It is the discursively and materially interrelated, socially constructed configuration of a specific policy environment, negotiated between and beyond structures and actors that lies at the centre of this analysis. Instead of regarding policymaking as a procedural formality or technocratic operation, I discuss the agency of the institution as part of a discursive institutionalist perspective (Schmidt 2010). This understanding of institutions enables an analysis that considers institutions as structures and constructs that are made and remade by political actors (Schmidt 2010: 14). Consequently, taking institutions not (only) as rigid collective bodies containing ideas, memories and rules, discursive institutionalism enables us to investigate the interpersonal and interorganizational exchanges of resources between the cultural administration and the *Koalition der Freien Szene*. Also, the interactions between the different concerned administrative departments will play a role. Finally, taking into account strategic-relational approaches such as Jessop (2006) and others like Neil Brenner, the relationality and reflexivity of policymaking is central to my investigation.

The Berlin City Tax debate – an agonistic policy arena

'Politics is first of all a matter of finding and defining the appropriate setting in which to stage the discursive exchange. These "sites" of discursive exchange have an influence on what can be said meaningfully and with influence' (Hajer 2003: 96). Borrowing from this image, in the following, I discuss the Berlin case of *Koalition der Freien Szene*, focusing on the process of negotiation on the first distribution of City Tax funds, generated since the introduction of the tax in 2014 and realized in 2015 (CityTax15).

Koalition der Freien Szene has, since its foundation in 2012, gained great attention with its critique of the apparent imbalance of Berlin's cultural funding system, attributing roughly 95 per cent of the cultural budget to fund art institutions such as Berlin's three operas, theatres or public libraries, while only 5 per cent of the overall funding volume goes to individual artists and projects, mostly cultural workers from the independent scene.[4] With its main policy document, the 'Ten-Point-Plan', the *Koalition* identified the structural underfunding of the independent scene and requested an additional €18 million for all artistic genres of the independent scene (*Koalition der Freien Szene* 2012). In the context of the introduction of the City Tax, the group demanded a redistribution of a significant share of the incoming monies to the independent cultural scenes. To strengthen its claim, in the fall of 2013, *Koalition* launched a public campaign (Landau 2015) and spoke to many members of parliament to convince them to vote in favour of directing the incoming monies to the independent scene. Throughout the budgetary negotiations for 2014/15, the initial propositions (ranging from directing all City Tax income to the independent scene to partitioning half of the income between three groups, namely the independent cultural scene, tourism and sports-related initiatives and projects) were heavily reduced by the budgetary committee. Contrary to the initial plan, €25 million out of the generated revenue of €29.4 million (2014) were subtracted for general budgetary consolidation purposes. The remaining €4.4 million, the so-called 'excessive incomes', were equally distributed amongst the three groups. In total, €1.38 million were available for arts and culture-related projects in 2015.

A legal plea against the introduction of the City Tax had been put forward by Germany's main hotel lobby organization, which was rejected in early June 2015. After the court decision, the City Tax income from 2014 became available to be distributed until the end of 2015 (Senatsverwaltung für Finanzen 2015b). This accelerated the collaborative process between the cultural administration and the *Koalition* towards designing a distribution mechanism for the available monies. A dialogue about distribution began in early 2015 exchanging suggestions and proposals articulated by both sides.

With the help of document analysis and interviews with *Koalition* spokespeople and employees from the cultural administration, in the following section, I explain how the CityTax15 process resulted into *Arbeits- und Recherchestipendien* (working and research grants – ARS; Table 5.2) for the genres visual arts, performing arts, music (new music/sound art), jazz and literature/poetry. Certainly, the effects of CityTax15 go beyond this one materialized policy output. Even though the following depiction of the process is anchored in empirical data, collected during as well as shortly after the negotiations in 2015, the given portrayal cannot claim to present an absolute depiction of the process. This is due to the general interpretational space of working with data sources, but also owed to the difficulty of reconstructing a process which took place mostly in non-public and non-accessible communicative settings. Using the policy outcome as an indicator of a concrete policy output, produced in the newly constituted network between the cultural administration and *Koalition*, I illustrate the emergence of the policy network via the exchange of various resources and the conditions and pressures under which these resource exchanges took place. Due to spatial constraints, I will not elaborate on other decisive dimensions of the CityTax15 process, such as the condition of in(ter)dependence between the two policy stakeholders or forms of acquiring and maintaining legitimacy through the provision of specific knowledge, attribution of symbolical power, reduction of insecurity, trust or other factors.

Negotiating distribution: Independent scene and cultural administration working together

The Berlin Budget 2014/15 was passed in December 2013 directing only excessive incomes of the City Tax to arts and culture. This outcome came to be known as the 'City Tax Lie' (*Koalition der Freien Szene* 2016). Shortly after the Budget had passed, *Koalition der Freien Szene* proposed the introduction of *Freier Kulturfonds Berlin* (Free Cultural Fund Berlin) with a funding volume of €5 million. This fund was designed to address the most urgent funding imbalances and was thought to keep up the pressure on politicians and the public after the dissatisfactory outcome of the Budget 2014/15 (*Koalition der Freien Szene* 2013). The idea of the fund somewhat resonated with public discourse, for example, when the opposition party, Die Linke, promoted the idea in a parliament plenary session in March 2015 (*Abgeordnetenhaus Berlin* 2015). Without further pushing towards realization, *Koalition* published the *Sofortprogramm* (Immediate Program) in December 2014, requesting a total volume of €3.98 million for supporting independent cultural production. Main demands included an increase of working and research grants and increasing resources for project funding (*Koalition der Freien Szene* 2014, see Table 5.1). *Sofortprogramm* shows the *Koalition*'s own prioritization of the most urgent funding gaps, and what, according to them, needed to be done first.

The visual arts organization *Berufsverband Bildender Künstler Berlin*, which also delegates spokespeople to the *Koalition*, commented that the City Tax process 'merely constituted an improvised contribution to the distribution of funds that were available at short-term, which should ideally be expertise-driven and use-oriented. This increase in production grants does by no means replace the substantial change of funding structures and distribution procedures requested by *Koalition*' (bbk 2015, translation: FL). Overall, *Sofortprogramm* provided the first collective quantified suggestion by *Koalition* of how to spend and invest potential monies from the excessive City Tax revenue of 2014. Developed from the core demands put forward in the 'Ten-Point-Plan', *Sofortprogramm* was taken to be its condensed version and provided a starting point for the concrete distributive negotiation between the cultural administration and the *Koalition*.

The first draft from the cultural administration regarding the distribution of the available funds, referred to as *Expressverfahren* (Express Procedure), was

Table 5.1 Funding gaps of Koalition der Freien Szene

Sparte		Betrag in Mio. €	für
Bildende Kunst		1,00	Zeitstipendien
Darstellende Kunst		1,00	Honoraruntergrenzen in der Einzelprojektförderung
		0,04	Geschäftsstelle LAFT
Tanz		0,10	Verdoppelung der Tanzstipendien
Musik	INM	0,32	Aufstockung der Projektförderung auf 0,5 Mio. Euro
	Jazz	0,20	Stipendien, Tourförderung, Studioprojekte (Einführung eines zweiten Fördertermins)
Literatur		0,18	Verdopplung der Autorenstipendien von 15 auf 30
		0,18	Aufstockung des Projektmittelfonds (Einführung eines zweiten Fördertermins und von Mindesthonoraren, Aufstockung der Maximalförderung auf 20.000 €)
		0,04	Geschäftsstelle, Erhebung zu Umfang und Beschaffenheit der freien Literaturszene
Projekträume		0,42	Verdreifachung der bisherigen Förderung

Source: Koalition der Freien Szene, Sofortprogramm.

introduced to *Koalition der Freien Szene* in a non-public meeting in early 2015. It suggested investing €880,000 for individual artist grants for all genres, an additional €210,000 for the project space prize, and the realization of an urban development project with a funding volume of €240,000. This project sought to address the highly politicized topic of lack of space for cultural and artistic purposes. Using a city-owned building for a one-time event in the fall of 2015, the idea was to create a trans-disciplinary public event and space, an 'Independent Scene Festival', where the independent scene would present and show itself. The whole festival would have been planned and carried out within less than a year, before the end of 2015. This proposition put forth by the cultural administration caused controversy amongst the members of the *Koalition*.[5] The suggestion for the urban development project, potentially to be carried out in collaboration with the urban development administration, was rejected by *Koalition der Freien Szene* in a meeting in the spring: 'You could never do something like that as *Koalition*, we are no cultural hosts, we are no cultural producers, we do not work on the notion of art itself or with art, we do cultural politics' (Interview KFS 2015). The idea was assessed by the cultural administration as 'too risky' or 'potentially an aftertaste of K2' (Interview SKA 2015), an event that had been criticized for its lack of financial and personnel resources to create a long-term commitment and dialogue with the cultural scenes. *Koalition* did not want to take on the role of 'event host' (Interview KFS 2015), and saw that the project could not adequately be realized with the temporal and financial resources offered by the cultural administration. Finally, the idea was rejected, and the monies added to the already planned working and research grants. The final scheme of distribution for the working and research grants, announced via open calls since July 2015, is laid out in Table 5.2.

Table 5.2 shows that a considerable amount of the requests from *Sofortprogramm* were realized with slight adaptations. For example, the increased funding for project spaces was doubled instead of being tripled as requested by the *Koalition*. Regarding the group's initial suggestion to increase project funding for the independent music scene, new individual working and research grants for musicians were created (see Table 5.1). The former suggestion for more project funding could not be realized because the modalities for project funding would have demanded more administrative effort than providing working and research grants. Other demands such as the establishment of administrative or office structures for the independent literature scene or the performing arts scene were not considered at all. However, it needs to be borne in mind that parallel to this distribution process – the one-time distribution of limited resources under time pressure – the Budget 2016/17 was being discussed between the cultural administration and the *Koalition*. In a meeting in February 2015, the administration made clear that they would closely consider the 'Ten-Point-Plan' in their Senate Draft for the Budget 2016/17. The Senate Draft took up demands such as minimum payment obligations or artist fees and sought to establish those as perennial budgetary items. These demands were thus not considered in the CityTax15 debate as this would have doubled positions or requests.

Interestingly, the relationship and communication between the cultural and the financial administration also played a role in determining what was technically and administratively possible in the process of CityTax15. The exchange between the two administrative departments about the status quo of the availability of the City Tax revenue from 2014 has been described as 'nebulous and intransparent' by a mid-level cultural administrator who was actively engaged in the operational distribution of the grants (Interview SKA 2015). Thus, the fact that the funds were unblocked just a couple of months before they had to be spent until the end of 2015 influenced the realm of action for the cultural administration, and intensified the pressure to distribute the funds in a quick and/or unbureaucratic fashion.

Exchanging capacity and relational resources

After having looked at the 'process of mutual positioning' (Hajer 2003: 107) between the cultural administration's and the *Koalition*'s perspectives and their respective (re)constructions of the CityTax15 process, I will synthesize these dynamics of the policymaking process with regards to the exchange of resources. Notably, the resources available to both policy stakeholders are diverse and asymmetrical, as the former is a volunteer-run loose organization with specific knowledge stemming from artistic practice, while the latter is a bureaucratically

Table 5.2 Arbeits – und Recherchestipendien (working and research grants) 2015, funded via City Tax income from 2014

Artistic Genre	Number of ARS	Number of Applicants	Amount of ARS (Euros)	Total Amount (Euros)	Distribution mechanism
Visual Arts	43 (at least 34 for individual artists, max. 9 for curators) 2015: 38 individual artists, 5 curators	1327	8000	344 000	new peer-reviewed jury
Project Spaces	7	n/a	30 000	210 000	existing peer-reviewed jury named list of substitutes
Performing Arts	min. 35 2015: 42	473	4000, 6000, 8000	284 000	existing jury
Jazz	min. 12 2015: 15	106	4000, 6000, 8000	96 000	existing jury
Literature	31	285	3 000	248 000	new peer-reviewed jury
Serious Music and Sound Art	19	188	3 000	152 000	new peer-reviewed jury
Total	minimum 140	2 379	maximum 8 000	1 334 000	

Source: author's depiction.

constrained entity with technocratic expertise and a public mandate for accountability.

For purposes of systemization, I subsume my observations under two types of resource exchange: First, scarce resources such as time, total funding volume and staff, which are referred to as 'capacity resources' and second, the exchange flows of communication, expectations, and legitimacy, denoted as 'relational resources', because they only exist in reciprocity or interaction between actors. Regarding the (material) capacity resources, the time pressure to distribute the funds and the scarce funding volume influenced the resource exchange between the cultural administration and *Koalition* regarding what was operationally possible. Certainly, some of the group's claims would have required much more expe/ ansive financial investments. Furthermore, the administration's staff capacities did not allow more labour-intense procedures to distribute the City Tax monies. Hence, as many working and research grants as possible were distributed through existing jury processes, which is an unbureaucratic distribution mechanism. The relational resources concerned the exchange of symbolic, communicative and cognitive properties between the cultural administration and *Koalition*. It is noteworthy that the frequency of communication increased significantly throughout the process of CityTax15; meetings, emails and phone calls were exchanged at a fast pace, also outside of business hours. With regards to their respective expectations, *Koalition der Freien Szene* might have wanted the cultural administration to realize all of their demands from *Sofortprogramm* (or even the more extensive 'Ten-Point-Plan'), which, in return, forced the administration with the challenge to manage – live up to, reject, modify and so on – these very expectations. With regards to legitimacy, *Koalition* was temporarily validated as policy entrepreneur by the cultural administration through the inclusion in the very process of CityTax15, while the cultural administration received legitimacy 'from the scene' because they developed a problem-oriented, useful new policy in collaboration with the artists themselves.

The governance arrangement inherent in CityTax15 is characterized by a high input legitimacy, that is, 'the process through which decisions are reached' (Fawcett and Daugjberg 2012: 202) and medium output legitimacy, that is, effective policy outputs. CityTax15 shows a high degree of input legitimacy as it included stakeholders in the policymaking process. Medium, not high output legitimacy was reached because much, but not all *Sofortprogramm*'s demands were realized in terms of effective policy outcomes. Going back to Rhodes' concept of a policy network, its definitional characteristics are evident and present in the CityTax15 collaboration: as a community of 'formal and informal institutional linkages between governmental and civil society actors' – in this case between the cultural administration and the *Koalition* – the two policy stakeholders gathered around the 'shared interest' of providing funding to producing artists (Rhodes 2007: 1244). The policy that emerged 'from the bargaining between the networks' members' (ibid.) is the policy outcome of the working and research grants. Thus, the policy network came into being because of and through the construction of a concrete policy output. With regards to the exchange of resources, the negotiation was based on a mutual understanding between the *Koalition* and the administration that the capacity resources were scarce. Regarding the relational resources, the cultural administration explicitly communicated what was possible and what could be done from their end, notably restrained by the interference and interdependence with other administrative departments, and what could not be realized due to constrained capacities. Through this interaction, both policy stakeholders also mutually experienced a state of in(ter)dependence.

Summarizing the interactions between the cultural administration and the independent art scene, three general conditions for collaboration changed in contrast to the years before: The particular actor constellation as well as the specific timing and the availability of financial means brought about an unprecedented monetary and symbolic increase in power for independent cultural producers in Berlin. First, the transdisciplinary *Koalition* provided a collective identity and offered a communicative opportunity towards the administration. Because potentially diverging opinions and sentiments from the genre-specific organizations are first discussed internally within the group and then carried towards the cultural administration, the *Koalition* facilitated the negotiations between the cultural administration and the formerly less tangible and addressable independent scene in a way. Second, the 'City Tax Lie' created a public sensitivity for the underfunded independent scene, so that the immediate and 'successful' distribution of CityTax15 was strongly desirable for the administration. The negative collective memory of the K2 cultural summit might have created additional pressure to enable a public encounter between the cultural administration and the independent scene, which would demonstrate a constructive collaboration. Thus, the timing and issue of CityTax15 revealed itself as a set of complementary needs and goals, as Wohlstetter *et al.* have termed it. Third, due to the disposable (however limited) of City Tax funds, the desire to establish a collaboration between the cultural administration and the independent scene was supported by concrete financial means that needed to be spent within a fixed timeframe. This had not been the case in earlier attempts to institutionalize collaboration between cultural administration and the independent scene.[6] The process of CityTax15 was centred on the solution-driven and time-limited distribution of actually available funds and operated on the basis of quantified demands from both the independent scene and the cultural administration.

Outlook: Institutionalizing the CityTax15 collaboration?

I have sketched explanatory factors of how and why a new policy network in Berlin's cultural policymaking context came into existence through the CityTax15 process. The emergence of this policy network shows how – in times of growing contestation of the narrative of the 'creative city' and structural challenges for urban governance – concrete and issue-related resource exchanges between different policy stakeholders can bring about policy collaboration and innovation. The CityTax15 process turned an idea into a new form of practice or the construction of a new institution (Innes and Booher 2003: 49). Although the institutional technical practices might not have been new, the particular communicative and collaborative practices between the cultural administration and the *Koalition* produced a novel material policy outcome within the temporary policy network.

With regards to the long-term stability and sustainability of this policy network, the most tangible and structure-changing indicator is the institutionalization of the distribution of City Tax. For 2016 and 2017, the so-called 'City Tax Fonds' will contain €3.5 million annually (out of the overall estimated City Tax income of over €47 million for 2015), having almost tripled in comparison to the monies available from CityTax15 (Der Regierende Bürgermeister von Berlin 2015), however, constituting just a fragment of the incoming funds. In a sense, CityTax15 has laid the foundation for the City Tax Fonds, which nonetheless differs in its conceptual structure and procedural distribution mechanisms. To assess and evaluate the evolution and long-term persistence of the policy network's agents and structures, we will have to look at the first material and symbolic-relational impacts of the City Tax Fonds. Future directions of the policy network will be developed reflexively along the lines of newly arising policy issues. The success and public reception of the first policy outcome, that is, the working and research grants, as a result of communicative and relational resource exchange between the cultural administration and *Koalition der Freien Szene*, will certainly affect the longevity of the policy network. In the end, the collaboration of CityTax15 has started to fill the institutional void of Berlin's cultural political terrain. We will see what resources will be exchanged in the future. Thanks to this start, maybe, the capacity resources – time, money and staff for the independent arts – can be increased towards improving the working conditions of Berlin's artists.

Notes

1 This chapter concentrates on the exchange of resources between the executive and a civil society actor (the *Koalition*), even though the legislative realm can certainly pertain to a policy network or governance arrangement as well. Due to the scope of my empirical investigation, encompassing interviews with members from the *Koalition* and employees from the cultural administration, the present analysis focuses on only the interaction between these two policy stakeholders.

2 Several contributions on Berlin's cultural and creative industries governance arrangements have appeared in the past years (e.g. Merkel 2012; Lange et al. 2008) as well as accounts from political science (Wostrak 2008) or cultural geography (Grésillon 2002) that provide historical or institutional overviews on the configuration of Berlin's cultural political field. Besides a cursory overview on cultural governance in an international comparative context (Merkel 2015), there have been few (empirical) accounts on current governance practices and mechanisms in Berlin's *artistic* field.

3 Although I operate with terms such as 'collaborative', I refrain from idealizing collaboration or deliberation or reifying consensus, as the former does not always and automatically produce democratic results. Regarding consensus, my approach stands in contrast to deliberative democracy approaches that assume that consensus can finally be reached through rational deliberation. My understanding of politics is rooted in agonism and assumes conflict to be an insurmountable dimension of human relations, so that consensus can only be established as contingent and temporary fixation of conflict (see Landau 2016b forthcoming).

4 The Cultural Budget for 2016/17 marks a notable increase of monies for the independent scene: In 2016, an additional 7.5 million euros will go into independent cultural

production, as well as 9.5 million euros in 2017. Furthermore, the stable budgetary item of 3.5 million euros, i.e. the City Tax Fonds, will be going towards the independent scene.

5 A point of controversy in the CityTax15 process was the introduction of working and research grants for curators, a funding gap highlighted by the cultural administration, as there are neither designated funding programs nor a collectivized lobby for curators in Berlin. The matter caused unease for *bbk* who wanted to make sure that the visual arts-designated working and research grants would go to individual practicing artists. After some back and forth, the issue was resolved by setting a maximum number of 9 out of 43 working and research grants to go towards curators.

6 For example, in the *Long-Term, Coordinated Dialogue with the Senate*, led by visual arts group *Haben und Brauchen*, no monies were available to be distributed in order to establish and enable collaboration.

References

Abgeordnetenhaus Berlin (2015) Freier Kulturfonds Berlin. Aktenzeichen: 17. Wahlperiode Drucksache 17/2148; Antrag der Fraktion Die Linke.

Baumgarten, B., P. Ulrich and P. Daphi (eds.) (2014) *Conceptualizing Culture in Social Movement Research*, Berlin: Palgrave Studies in European Political Sociology.

Benford, Robert D., D. A. Snow, H. J. Mc Cammon, L. Hewitt and S. Fitzgerald (2014) 'The emergence, development and future of the framing perspective: 25+ years since "frame alignment"'. *Mobilization – An International Journal* 19 (1): 23–45.

Berufsverband Bildender Künstler (bbk) (2015) City Tax Mittel 2014. Brief der Koalition der Freien Szene an den Kulturstaatssekretär Tim Renner und an die Mitglieder des Kulturausschusses im Abgeordnetenhaus zu Berlin; accessed 27 January 2016 at www.bbk-berlin.de/con/bbk/front_content.php?idart=3911&refId=199.

Börzel, T. (1998) 'Organizing Babylon – On the different conceptions of policy networks'. *Public Administration* 76 (1): 253–273.

Der Regierende Bürgermeister von Berlin (2015) Rote Nummer 2472 B. Kapitel 0310 – Kulturelle Angelegenheiten Titel 68627 – Zuschüsse für besondere kulturelle touristische und sportbez. Abgeordnetenhaus Berlin. Berlin (Rote Nummer 2472 B); accessed 18 January 2016 at www.parlament-berlin.de/ados/17/Haupt/vorgang/h17-2472.B-v.pdf.

DiGaetano, A. and E. Strom (2003) 'Comparative urban governance. An integrated approach'. *Urban Affairs Review* 38 (3): 356–395.

Dowding, K. (1995) 'Model or metaphor? A critical review of the policy network approach'. *Political Studies* 43 (1): 136–158.

Fawcett, P. and C. Daugbjerg (2012) 'Explaining governance outcomes. Epistemology, network governance and policy network analysis'. *Political Studies Review* 10 (2): 195–207.

Fischer, F. (2003) *Reframing Public Policy: Discursive Politics and Deliberative Practices*, Oxford, New York: Oxford University Press.

Fischer, F. and H. Gottweis (2012) *The Argumentative Turn Revisited; Public Policy as Communicative Practice*, Durham, NC: Duke University Press.

Fung, A., E. Olin Wright and R. Abers (2003) *Deepening Democracy; Institutional Innovations in Empowered Participatory Governance*, London, New York: Verso.

Goodin, R. E., M. Moran and M. Rein (2008) *The Oxford Handbook of Public Policy*, Oxford, UK: Oxford University Press.

Gottweis, H. (2003) 'Theoretical strategies of poststructuralist policy analysis: Towards an analytics of government', in Hajer, M. A. and H. Wagenaar (eds.) *Deliberative*

Policy Analysis: Understanding Governance in the Network Society, Cambridge and New York: Cambridge University Press: 247–266.

Grésillon, B. (2002) *Berlin: Métropole Culturelle*, Paris and Berlin (Mappemonde).

Hajer, M. A. (2003) 'A frame in the fields: Policymaking and the reinvention of politics', in Hajer, M. A. and Wagenaar, H. (eds.) *Deliberative Policy Analysis: Understanding Governance in the Network Society*, Cambridge and New York: Cambridge University Press: 88–113.

Hajer, M. A. and Wagenaar, H. (2003) *Deliberative Policy Analysis: Understanding Governance in the Network Society*, Cambridge and New York: Cambridge University Press.

Haunss, S. (2004) *Identität in Bewegung. Prozesse kollektiver Identität bei den Autonomen und in der Schwulenbewegung*; Wiesbaden: VS Verlag für Sozialwissenschaften.

Hay, C. (2002) *Political Analysis*, Houndmills, Basingstoke and Hampshire, New York: Palgrave.

Hay, C. and D. Richards (2000) 'The tangled webs of Westminster and Whitehall: The discourse, strategy and practice of networking within the British core executive', *Public Administration* 78 (1): 1–28.

Innes, J. E. and D. Booher (2003) 'Collaborative policymaking: Governance through dialogue', in Hajer, M. A. and Wagenaar, H. (eds.) *Deliberative Policy Analysis: Understanding Governance in the Network Society*, Cambridge and New York: Cambridge University Press: 33–60.

Interview Transcript (2015), Koalition der Freien Szene, September 2015.

Interview Transcript (2015), Senatskanzlei für Kulturelle Angelegenheiten, July 2015.

Jessop, B. (2006) *State Power: A Strategic-Relational Approach*, Cambridge, UK: Polity.

Kingdon, J. W. (2011) *Agendas, Alternatives, and Public Policies*, Boston, MA: Longman.

Koalition der Freien Szene (2012) Zehn Punkte für eine neue Förderpolitik 2012; accessed 27 January 2016 at www.berlinvisit.org/forderungen-zahlen-2.

Koalition der Freien Szene (2013) Vorschlag Freier Kulturfonds Berlin, December 2013; accessed 27 January 2016 at www.berlinvisit.org/vorschlag-freier-kulturfonds-berlin.

Koalition der Freien Szene (2014) *Sofortprogramm*, Koalition der Freien Szene: Berlin.

Koalition der Freien Szene (2016) Die City Tax Lüge, accessed 27 January 2016 at www. berlinvisit.org/die-city-tax-luege/.

Kucher, Katharina (2013) 'Öffentliche Kulturförderung der Freien Szene Berlin. Eine Politikfeldanalyse'. Unpublished Masters Thesis: 1–74.

Landau, F. (2015) 'Tagging the City: Berlin's independent scene rising for a City Tax for the arts', *edgeCONDITION Special Issue Placemaking* 1 (1): 118–21.

Landau, F. (2016a) 'Articulations in Berlin's independent art scene – on new collective actors in the art field', *International Journal of Sociology and Social Policy*, 36 (9/10); in print.

Landau, F. (2016b) 'Unpacking conflictual consensus in Berlin's cultural policy-making'. Manuscript under review for publication.

Lange, B., A. Kalandides, B. Stöber and H. A. Mieg (2008) 'Berlin's creative industries. Governing creativity?' *Industry and Innovation* 15 (5): 531–548.

Marsh, D. and R. A. W. Rhodes (1992) *Policy Networks in British Government*, Oxford, New York: Clarendon Press.

Marsh, D. and M. Smith (2000) 'Understanding policy networks: Towards a dialectical approach'. *Political Studies* 48 (1): 4–21.

McAdam, D., S. G. Tarrow and C. Tilly (2001) *Dynamics of Contention*, Cambridge, New York: Cambridge University Press.

Merkel, J. (2012) 'Creative Governance in Berlin'. In Anheier, H. K. and Y. R. Isar (eds.) *Cities, Cultural Policy and Governance*, London, Thousands Oaks, California: SAGE (The Cultures and Globalization Series, vol. 5): 160–166.

Merkel, J. (2015) Berliner Kulturpolitik in international vergleichender Perspektive; accessed 27 January 2016 at www.hertie-school.org/fileadmin/images/Downloads/ pressmaterial/kulturstudie/Hertie_School_Studie_Berliner_Kulturpolitik_in_interna- tional_vergleichender_Perspektive_WEB_Aufloesung.pdf.

Mintrom, M. (1997) 'Policy entrepreneurs and the diffusion of innovation'. *American Journal of Political Science* 41 (3): 738–770.

Moutte, C. (2005) *On the Political*, London, New York: Routledge.

Peck, J. and N. Theodore (2012) 'Follow the policy: A distended case approach'. *Environment and Planning A* 44: 21–30.

Pierre, J. (2005) 'Comparative urban governance: Uncovering complex causalities'. *Urban Affairs Review* 40 (4): 446–462.

Pierre, J. and G. Peters (2000) *Governance, Politics, and the State*, New York: St. Martin's Press.

Rhodes, R. A. W. (2008) 'Policy network analysis', in Goodin, R. E., M. Moran and M. Rein (eds.) *The Oxford Handbook of Public Policy*, Oxford: Oxford University Press: 425–448.

Rhodes, R. A. W. (2007) 'Understanding governance: Ten years on'. *Organization Studies* 28 (8): 1243–1264.

Sabatier, P. A. (2014) *Theories of the Policy Process*, Boulder, CO: Westview Press.

Senatskanzlei Kulturelle Angelegenheiten (2015) Mit dem neuen Haushalt 2016 / 2017 gibt es ein neues Miteinander zwischen Stadt und Freier Szene; accessed 27 January 2016 at www. berlin.de/sen/kultur/aktuelles/pressemitteilungen/2015/pressemitteilung.420975.php.

Senatsverwaltung für Finanzen (2015a) FAQ zur Übernachtungsteuer (City Tax); accessed 15 July 2016 at www.berlin.de/sen/finanzen/steuern/informationen-fuer-steuerzahler-/ faq-steuern/artikel.57911.php.

Senatsverwaltung für Finanzen (2015b) Nach Entscheidung des Finanzgerichts zu City Tax: Finanzverwaltung gibt Mittel für Kultur, Sport und Tourismus frei – Berlin. de. Senatsverwaltung für Finanzen; accessed 27 January 2016 at www.berlin.de/sen/ finanzen/presse/pressemitteilungen/pressemitteilung.325166.php.

Schmidt, V. A. (2010) 'Taking ideas and discourse seriously: Explaining change through discursive institutionalism as the fourth "new institutionalism"' *European Political Science Review* 2 (1): 1–25.

Spiegel Online (2014) 'Arm, aber sexy: Wowereits beste Sprüche'. In *Spiegel Online* 2014, 2014; accessed 27 January 2016 at www.spiegel.de/politik/deutschland/klaus- wowereit-seine-besten-sprueche-a-988169.html.

Tarrow, S. G. (2011) *Power in Movement: Social Movements and Contentious Politics*, Cambridge and New York: Cambridge University Press.

Woddis, J. (2013) 'Arts practitioners in the cultural policy process: Spear-carriers or speak- ing parts?' *International Journal of Cultural Policy* 20 (4): 496–512.

Wohlstetter, P., J. Smith and C. L. Malloy (2005) 'Strategic alliances in action towards a theory of evolution'. *The Policy Studies Journal* 33 (4): 419–442.

Wostrak, A. (2008) *Kooperative Kulturpolitik. Strategien für ein Netzwerk zwischen Kultur und Politik in Berlin*; Frankfurt am Main, New York: Peter Lang (Studien zur Kulturpolitik, Bd. 6).

Yanow, D. (2003) 'Accessing local knowledge' in Hajer, M. A. and Wagenaar, H. (eds.) *Deliberative Policy Analysis: Understanding Governance in the Network Society*, Cambridge and New York: Cambridge University Press, 228–247.

6 Alternative art schools in London

Contested space and the emergence of new modes of learning in practice

Silvie Jacobi

Introduction

With the tripling of tuition fees for higher education (HE) in England and Wales in 2012, and the gradual absorption of art schools into the context of the neoliberal university, a considerable number of artist-led training opportunities emerged across the United Kingdom in response to the need for affordable art school education. This research focuses on three different models for alternative art schools that have emerged in London since 2012. The chapter establishes a brief context of art school education in the United Kingdom and then looks at London's urban context to highlight the conditions that have amplified the contestation of space for artistic production.

The increasing financialisation of art schools aligned with urban struggles build the framework around which the findings are interpreted. Empirical data was collected through interviews with the organisers and founders of the schools, outlining both ideological and logistical reasons for setting up the schools, which includes the important aspect of artist's access to space for a studio-based art education. The availability of space was a crucial factor for the organisation and format of education offered, with one case study questioning the extent to which physical space is required at all.

As the case studies represent a widening of professional profiles and resilience capacities of artists, the chapter focuses on artists' role in appropriating space and engaging a diverse public including the local community. Each operational model of the schools is fundamentally structured around the nature of the artistic practices the schools explore (in conversation with the space available). This ranges from traditional studio-based work to socially engaged practices, with the latter providing more opportunities to attract public funding through the community focus. To run the schools as cost-neutral as possible, they are based at sites of looming regeneration, situated at or nearby large social housing estates. The initiatives respond to the need for accessible and flexible training for visual artists not necessarily given by existing HE fine art departments due to high tuition fees and targeted learning outcomes. Along with the provision of a learning environment the alternative art schools fulfil the need for affordable studio space and an opportunity to network.

UK art schools

British art schools were traditionally seen as an accessible alternative to the high cultural university,[1] providing a diversity of people especially those from a working-class background with aesthetic and craft-based technical training (Banks and Oakley 2016). By the early 1960s, art schools offered further education, short courses and degree-equivalent diplomas as independent institutions funded by Local Education Authorities, which on top of supporting free tuition offered maintenance grants to students. This provision met some of the demand by industry, which required professionals in craft trades such as textiles, ceramics and other forms of design industrial manufacture (ibid: 3).

At the heart of art schools, as Banks and Oakley underline (ibid: 6), were the cultivation of human capital and social upward mobility as a state-authorised opportunity to exercise social democracy and radical ideas that offered innovation. Free education and the independence of schools ensured that artistic pedagogies were open-ended yet material-specific enough to allow for a balance between technical skill and artistic freedom. Yet tensions increased between the needs of art and industry, with an increasing demand for artistic freedom aligned with the conceptualisation of art. Along with the de-skilling of art in a post-Duchampian era (Madoff 2009), students demanded more theory and research as part of their art and design education as well as a break up of hierarchical structures between students and teachers (Tickner 2008). These demands were to some extent aligned with the 1960 Coldstream Report,[2] commissioned by the National Advisory Council on Art Education, which recommended the introduction of theory elements to art school education. This report, however, fed into the rolling out of accreditation mechanisms and timetabling of art school education, which decreased certain freedoms such as the open-endedness that is core to artistic pedagogies (especially in fine art).

By the mid-1970s more universities began to offer art and design degrees, and art schools were being absorbed into polytechnics. Another wave of rationalization of art school education happened during the 1990s, under Tony Blair's New Labour, with a vast increase of courses in the arts and design arena (Oakley, Sperry and Pratt 2008) to educate the human capital ('bohemian graduates') for a growing cultural and creative industries sector (Comunian, Faggian and Li 2010). Under Labour, tuition fees were first introduced in 1998 across the United Kingdom as a means of funding tuition to undergraduate and postgraduate certificate students at universities, with students being required to pay up to £1,000 a year for tuition. The fees then gradually increased until 2012 when tuition fees tripled to a level £9,000 per year for undergraduate degrees. This was aligned with the then Tory-Liberal Democrat coalition government's stated goal to fight the United Kingdom's deficit, which saw the decrease of public funding for education as well as the arts. For art schools this meant a demeaning of their liberal purpose, through monetising an education that crucially relies on open-ended content and method of teaching, with outcomes not quantifiable and instead value placed on critical thought. Art schools require workshops where one can make physical things alongside having unstructured studio time, which most universities are reluctant to accommodate given the cost of space in London and a trend towards providing digital facilities at the expense of material practice. American feminist artist Judy Chicago (2014) questions the high levels of fees paid for an education that is in flux, and calls for a renewed focus on training technical skills needed for artistic practice as well as professional skills such as working with curators, writing critically within a visual arts context or managing self-employment to sustain a lifelong practice, without immediately being pushed towards success in the art market.

Looking at these unique qualities of art-school education, one can question whether art departments at university can still be referred to as 'art schools', as the institutional autonomy both in terms of having a building that can accommodate artistic practices as well as open-endedness in teaching is not necessarily given within the context of university's increasing commercial culture (Ivison and Vandeputte 2013). I would argue this goes along with confusion about which subjects belong in an art-school education, and which have seen a shift from traditional art and design subjects to subjects related to the creative industries aligned with communications, digital technologies and arts management, for example. Banks and Oakley (2016) conclude that 'questions of access, equality and representation betray how far we have come from the cultural and social milieu [of art schools]'. In British universities, access to education is monetized and the majority of students take on the burden of debt through loans, with the risk that they may never be able to pay them off due to precarious employment in the cultural industries. As Chatterton (2000: 178) concluded, the 'critical and analytical functions of universities in the community may be subject to erosion' if the university (and art schools) are no longer public resources and no longer reflect the needs and knowledge of society.

In 2012, for the first time, applications to art and design courses dropped (Asquith 2014), which seems to show the immediate effect of HE's financialisation bringing along a much more calculated approach to careers along with a general devaluation of art and design arising from government cuts. There has been repeated mainstream liberal and art world press coverage on the rise of alternative art schools in London and other parts of the United Kingdom, while reports also emerged about art student protests such as at Central Saint Martins.[3] Artist development organisations, university fine art departments and artist studio providers across London recognize the increasing pressure of sustaining an artistic livelihood in London. This has culminated in a range of events and conferences in 2015 around the issue of the survival of art schools entitled 'System Failure Talks' (Artquest 2014).

London is changing

Evidence from the campaign design project 'London is changing' (Londonis changing.org 2016), set up by a lecturer in communication design at Central Saint Martins, depicts the many accounts of relocation to, from and within London as a measure of the direct impact of recent economic and policy changes to the culture and diversity of London. Many of these voices captured a feeling of being forced out, while others stress how the city is 'eating itself' in the face of cultural diversity being lost (Moore 2015). Undoubtedly, there is an intensifying wave of super-gentrification (Butler and Lees 2006) that pushes lower-income Londoners further out, while inner-London and other desirable areas of Greater London experience an influx of wealthier (often global) elites. Media reports are quick in blaming those elites for the rush into buying London properties as safe assets or for consuming an elite higher education. It is, however, the political environment that to some extent allows these elites to influence housing and planning in such a way that an affluent luxury housing sector could emerge in London and other parts of the United Kingdom, arguably having a negative impact on the supply and affordability of housing for the rest of the market (Paris 2013; Hamnett 2009). Based on Land Registry Data on house prices, Reades (2014) illustrates the significant expansion of unaffordable housing between 1997 and 2012. He also points out how even the formerly affordable East End 'is now at best, something of a stretch for most households' (ibid: 338), which leaves vast amounts of low to middle income earners locked out of the London housing market. The student accommodation market has also undergone a bizarre 'luxurification', often only by an increased monetary value while many halls are in dire need of refurbishment. This just briefly touches a few dimensions on Britain's housing crisis with London being at the epicentre. While housing is an issue, it puts pressure on spaces for industrial production, which in relation to the cultural industries has meant that spaces of consumption put those of production at risk (Pratt 2009).

All these scenarios paired with decreasing public funding for the arts provide an intense amount of pressure on the survival of artists in London along with artist studio providers and the cultural economy they rely on. A significant number of art world talks (e.g. Frieze 'Off centre: Can artists still afford to live in London'[4]) have recently picked up on the paranoia of artists' fears of being displaced from London. This was set on fire by the Greater London Authority (the administrative body for Greater London) who estimated that as many as 3,500 artists are likely to lose their places of work within the next five years, which is a loss of 30 per cent of the current provision (GLA 2014). While this report shows an anecdotal commitment to 'tactical interventions'[5] that are temporary and showcase artist's resilience capacity and creative adaptability, they betray the long-term needs of artists. New modes of ownership are urgently required to protect space from the forces of an unregulated property market. These inclusive ideals, however, require a different political environment, within which the city's social and cultural functions are protected from absorption into neoliberal markets.

Alternative art school modes

From staged 'utopias' to empirical insight

Until now, the rise of alternative art schools as reported through the press and art world media has not been discussed from a social science perspective. Hence, this research seeks to establish the modes of organization through which these art schools base their practices, and how these are determined through different funding and space provision approaches. This assumes a specific relationship between location and education, and this study highlights how these intersect in conversation.

This research is based on interviews with the founders and organisers of three schools with the aim of discussing their inception, ideologies and operation. This is crucial to articulate how location along with access to space influences provision of education and opportunities for engaging art-world audiences and the local community. I identified the initiatives based on personal knowledge of the art world in London and chose the cases based on their self-recognition as 'schools'.[6] Further data on student's projects and experience was then analysed through secondary data as provided on the schools' websites and reports written by students and staff. As a student from one of the schools stressed, alternative art schools are 'temporarily enacted utopias' (Artquest 2014: 13) and as such 'an idealisation that has come into being [which] can now be used to make visible ideologies that are commonly hidden from view through their common sense qualities' (ibid). By studying how alternative ideas have been brought alive, this research helps to analyse the potential repercussions of these on the wider system of art school provision in the United Kingdom.

Turps Art School

Turps Art School grew out of the initiative of professional painters and lecturers, who published an independent painting magazine called *Turps Banana*. The idea for the school bore out of the magazine as a way to extend its knowledge exchange platform. In 2012, when the rare opportunity arose to rent out a large physical space in Bermondsey, South East London, they decided to pioneer the project. With a year break to find another space that offered more permanence, Turps moved into Taplow House, located in the yet to be developed Aylesbury Estate in Elephant and Castle, South London. This opportunity arose in connection with artist studio provider, Artist Studio Company (ASC), having moved there. Built between 1967 and 1977, Aylesbury Estate became the symbol of a broken Britain under New Labour government in late 1990s (Lees 2014). Subsequently, the estate received the 'New Deal for Communities' status and residents voted for its regeneration. Yet under current development plans, residents will be rehoused, as it is feared, to peripheries of or outside of London. This has been the case with the neighbouring Heygate Estate (Heygatewashome.org 2016). Whereas Heygate has already been demolished and new-builds are being set up, Aylesbury will be taken down in several stages, with artists using empty parts of the estate on an up to ten-year lease.

Turps' structure is based on a network ('peers') of painters, who share a communal studio space across three formerly commercial and public-use units on the second floor of the housing block. Students are selected on a basis of their portfolio and admission is granted through a fee payment of currently £6,000 that covers tuition and studio for one year. While this is cheaper than formal HE, which is currently at around £9,000 for undergraduate degrees, the financialisation of a course that does not lead to a qualification is problematic. Financialisation of art education is common practice in many parts of the world (e.g. the United

Kingdom, the United States and Singapore) where art education is expensive and to some extent exclusive if no student loans are available – as is the case for post-graduate programmes in the United Kingdom. In contrast with this, continental-European schools are still publicly funded but highly competitive and hard to enter. In that light, Turps satisfies the need for access to art education at the inter-section between affordability and less competition. The cheaper HE offer makes a distinct business case for the school, which the team argues currently does not receive any public funding for their educational arm. The gallery space at Taplow House is the only element at Turps with public support, as it involves a public educational programme.

> The reason I think we can do that [run the school], is because I think mainstream education is private education when you're paying £9000 a year – there's no difference. So my whole starting point is, if you can make it cheaper and you can make it more succinct and more direct, then you have a business that should thrive.
>
> (Group interview: Founders/Organisers, 2015)

Whether the gallery with its public funding can be separated from the school is debatable, as it has in the past received Arts Council England funding for educational activities such as artist talks. What Turps considers a postgraduate programme is financially viable through affordable rent and self-generated income from students not just paying for space but also for the reputation of mentors associated with the school. These are lecturers from across art schools in London. To provide a much-needed alternative with value for money was a response to a 'climate where there is growing disquiet about the quality of painting tuition, inadequate tutorial input and isolation in traditional studio set ups' (Turps Banana 2016). The emphasis of the school is a return to the studio and focus on painting practice away from calcu-lated outcomes and students pursuing a 'performative act of being artists' (Group Interview: Founders/Organisers, 2015). The school aims to reengage professional painters with their practice in a community that considers any pastoral care as part of simple human communication and not a calculated provision of HE.

> As artists we give extra, because it is our life.
>
> (Group interview: Founders/Organisers, 2015)

However, this 'giving extra' seems restricted to the immediate network and leads primarily to satisfying the agenda of providing opportunities for their peers and providing an environment for focused artistic practice.

> We don't get any government money to support. So we don't owe anything back to the community, but we behave in a manner that is noble by our prac-tice. And therefore everyone here should be a good citizen of the estate, and that comes with the participation on the course.
>
> (Group interview: Founders/Organisers, 2015)

Despite receiving some public funding for the gallery, the team felt that a potential social engagement role within the housing estate they are based at would be prescribed artificially and would wrongly assume that all artistic practices can be instrumentalised for non-artistic goals. Due to painting being studio-based and their mission providing a 'back to the studio/back to practice' environment, other artistic practices can cater much more for social engagement. Turps sees their role rather in being good neighbours:

> The people who live here are here because they can afford to live here. We're all here because this is where we can afford to be. It doesn't mean that the people upstairs or us wouldn't like to live in better conditions, like with bespoke studio spaces and so on. But it's just this happens to be what is possible, it happens to be what works.
>
> (Group interview: Founders/Organisers, 2015)

The school acknowledges their indirect role in regeneration as the deal with the developers of the estate led by Southwark Council was negotiated on the basis that they would 'bring people to an otherwise unpeopled space' (Group interview: Founders/Organisers, 2015). This hints at Bourdieu's (2011) conception of artists representing a specific class through cultural capital and how their presence in a formerly working-class and African-Caribbean area of London is an indicator of gentrification. In light of looming redevelopment, which the artists consciously make use of, the interviewees highlighted how the ten-year lease, which they considered long term makes the business case possible. Within this timeframe, they point out how any investment (e.g. studio fittings, marketing, unpaid time of personal engagement) can be recovered.

Open School East (OSE)

In 2013, a team of visual artists and curators involved in projects associated with London's Cultural Olympiad in 2012, came across the Rose Lipman Library, a former children's library in Homerton, East London, which prompted them to set up an alternative art school. The library is part of De Beauvoir Town, which is a low-rise area of mostly Victorian housing in Hackney that saw a period of intense, localised and collective action against demolition of Victorian streets in the 1970s. The estate consists mainly of tower blocks with the Rose Lipman Library building opened in 1975 as community centre and children's library. Hackney Council is currently considering the site for redevelopment into mixed use (residential, commercial and community), but this does not amount to a specific plan[7].

Open School East's (OSE's) organisational model is based on leveraging public funding for running a student-led community outreach programme, which acts as an extended learning environment for artists involved. The school recruits around fifteen associates (students) each year, which are mainly visual artists and in some cases other practitioners or researchers wishing to explore new media and social subject matter (Bourriaud 1998). The application process includes a portfolio and expression of interest not necessarily aimed just at socially engaged practitioners, but at artists who are 'socially-minded'. Associates enjoy free studio provision and mentor-led training made possible through public funding. However, in exchange, associates are expected to actively curate the public programme, which includes putting on engagement and learning events as well as longer term projects involving the local community. Seed funding for the public programme was initially provided in 2012/13 through the Barbican Centre in collaboration with public art agency CREATE.[8]

> The studios and education are free because there is an exchange. Whether you like it or not, a lot of work has gone into raising the money to make this happen. You know we have to deliver things associated with that money. It's about everyone understanding their time here is partly what is being delivered, but also that the remit of the school is to work with lots of different kinds of voices and to be a resource for these.
>
> (Interview: Co-founder)

Additionally, the directors at OSE build on their professional links in the art world, which prove invaluable for securing funding (e.g. through benefit auctions and patronage) and sourcing renowned lecturers from London art schools as mentors to staff tutorials and critique session ('crits').

> The case of OSE can offer valuable insights into social and power relations established within a non-fee paying art school, and how they affect our experiences as students.
>
> (Student feedback: Artquest 2014)

While the learning environment was perceived as a comfortable space in which difficult artistic questions could be faced more consequentially through a group setting (Artquest 2014), it was problematized that practices of associates were instrumentalised and consumed through the public programme, which as one associate put it 'might … internalise the pressure of our role in the school's survival' (Artquest 2014: 13). Taking aside these internal tensions, the scope of initiatives emerging from the school suggests that OSE is a springboard for community activities and longer term activist initiatives. The range of people OSE attracts to the space highlights that it is a public resource not just for artists and local community groups, but also for a wider community of people wishing to take part in cultural activities.

> A group of very radical academics from [a university in London] who are perhaps disillusioned with how universities have changed in the last 10 years felt that this was the kind of space that had a lot mutually going on. So we can help them to reach a much more genuine and varied reach of people, the actual public, and they can teach in a much more inspiring and interesting way.
>
> (Interview: Co-founder, 2015)

The associates' public projects, for example, evolve around localised issues on urban regeneration, both identifying the history of the De Beauvoir estate and linking this with current visions for regeneration and housing. Output of many of the projects that associates develop are lecture-style events, social gatherings based around food and music, which act as a platform for building relationships, performances, artist talks, research and education opportunities for local residents. These findings suggest an important role for OSE in promoting engagement with local residents and the wider public in knowledge exchange, which provides them with a heightened sense of connectivity with place and people outside their comfort zone. While OSE provides opportunities for the public from within the art world as well as working with communities at risk of marginalisation, the organisation sees itself faced with constant survival struggle in the face of gentrification:

> One big question we have at the moment is: "Do we stay in London?" We're feeling in our own lives that it's a difficult city to live in if we want to work the ways we do. We want to office the people who can't necessarily get [education] through other means. Well, if they can't be here they can't access it anyways. The reasons these provisions are appearing is because they are necessary I think.
>
> (Interview: Co-founder, 2015)

One way the school has identified to scale up is to move to another site with easier ground-floor access to attract more footfall, as well as space for affordable studio provision for former associates as a way to diversify income.

The School of the Damned

While the other two cases evolve their operational model around a fixed studio space, School of the Damned (SOTD) is a UK-based network of mostly fine art graduates who meet once a month at an artist-led space above a pub in Euston, Central London. Even though schools are generally associated with physical spaces, the SOTD group are very determined in defining themselves as school despite the lack of a building. Most students have individual studio spaces across the United Kingdom. In some cases, their practice is not even bound to a studio space. The focus of the school became the coming together as practicing artists and engaging in critical conversation of their work:

> We're there to learn and better ourselves, which is what you do at school. Just because we don't have a building ... I think it is still important to keep us in relation to art school and to be seen as the same but different. You don't just want to call what we do with a random name. It would be something totally different. People wouldn't necessarily understand what we did.
>
> (Interview: Student member, 2015)

The function room in which SOTD meet is run by local artists who negotiated a 'deal' with the pub owner to put on exhibitions and talks, most of them aligned with other activist movements and self-organized academic groups, with the aim to increase footfall to an otherwise tired pub. SOTD itself is based around what the group identifies as a 'labour exchange' model through which they attempt to withdraw themselves from the monetary system and test what situations and opportunities can emerge through holding onto this 'illusion'. Labour exchange means that value is created through social interaction and bartering, that is, using an exhibition space for free in exchange for artist's DIY skills to fix the space up.

> I think the SOTD is more politically charged than most of the other alternative art schools. So there is this aspect of protest within it. And quite a lot of us are very aware of wider political situations and framing ourselves within that debate. And the aesthetics of the SOTD and the manifesto, are very much in opposition of marketization of education and finance. So it obviously attracts people who are more aware of those things, I guess.
>
> (Interview: Student member, 2015)

Prospective students are recruited by open call and portfolio, and current students are responsible for the selection and hand over (of digital resources and knowledge) to the new cohort. As the artists are not just based in London but also in other parts of the United Kingdom, this provides opportunities to pull in everyone's resources – the art market in London and the more affordable exhibition spaces outside London. Also, the continuity of having a network that is run for one year has been perceived invaluable in forming relationships, which to some extent led to residency exchanges and members sharing studio spaces.

The school operates without any formal hierarchy and its members are as actively involved as their other professional commitments with full- or part-time work allow. Mentors who are invited to lead critique seminars are referred to as 'guests' rather than assigning a hierarchical position to them, which is an example of how language shapes 'utopian' realities. The school has been running since 2012 and cohort after cohort are building a larger network of graduates with a common knowledge base, which sends a strong message around alternative arts education being perhaps equally 'efficient' as its monetised equivalent. The longer the school carries on and because everyone has an active role in running it, the group believes that people will stay engaged and this may have a signalling impact to leverage more space opportunities and teaching time through guest lecturers as the network of graduates grows.

Conclusion

The brief discussion of the three cases has aided a first introduction to different modes of art-school development outside formal HE. This provides not just examples for how an art school could be organised differently, but most importantly shows the complex needs towards space provision engendered by the schools' practice specialisms. Within art's expanded field[9] (Krauss 1979), artistic practices are not bound to the studio as a production environment and stretches the boundaries of their material specificity. Yet, as the case studies have shown, painting requires studio space whereas the production of relational art (Bourriaud 1998) as in the case of OSE, is situated beyond the studio and can be embedded in local communities. Whereas Turps builds on subject matter that is removed from the immediate community context and is more inward looking because of its non-relational focus, OSE (and SOTD to some extent) internalise their location context in the development of subjects based on everyday experiences. By looking at the evidence of Turps, painting as a studio-based practice requires a structured and permanent environment within which practice, and the pedagogies surrounding it, can evolve. The school is charging tuition fees to provide for this environment. OSE's model in contrast shows how social engagement practice can extract exchange value to secure free education for its associates, but at the expense of some of the student's artistic autonomy. OSE has become a cultural resource and asset for the local community and wider publics interested in knowledge exchange. There is clear evidence from this study that the discussion of space has an impact on how art-school education is delivered, whereas also the distinct modes of production (despite the expanded field) mean different organisational opportunities and challenges for the school.

Artistic needs and uses have become ever more integrated in one space – linking education, artistic practice and social engagement. This illustrates how visual artists are now required to build an 'all-in-one' case to secure space and funding, which can perhaps be understood also as a way to legitimise their presence in the urban regeneration hierarchy in highly spatially contested cities. But if the schools cannot acquire ownership over a site, they remain instrumentalized in the urban regeneration context. Despite the cultural capital of the school's members, there is an inherent danger of their displacement through redevelopment of the site. Alongside this, the extended community that the schools serve is subject to gentrification, which is another factor that puts the schools' activities at risk in the long term. Despite these issues, alternative art schools in London lead the way in providing space for artistic production in a highly contested city, including space for the re-production of artistic discourse and the development of artistic human capital. It can be questioned to what extend the schools have a transformative impact on individual students and the local community as applicable. The timeframe for conducting fieldwork for this study (October to December 2015) was too short to investigate the expectations against individual transformative outcomes of those involved. However, in terms of transformative outcomes on the larger HE system, the schools have a signalling impact through highlighting the contrast between what they offer with an equivalent overpriced art school experience in formal HE. In particular, the activities at SOTD have internalised the extent to which the value of formal art schooling became extorted. The group's monthly activities parody the existing HE fine art offer by showing how effortless and affordable it is to utilise existing knowledge and social capital to provide a learning platform based on exchange. The aspects of Turps charging fees and OSE trading artist's social engagement in exchange for public funding can be understood as an internalisation of neo-liberal struggles – an unavoidable compromise for the schools to be possible at all. While this chapter presented a London-centric example of artists' resilience, it sketches new and practical avenues for resisting neo-liberal HE, which can be explored in other national and regional contexts.

Notes

1   British higher education has a binary system that consists of traditional research-intensive universities serving an elite and reproducing 'high culture' and former poly-technics with strong vocational focus. Despite polytechnics acquiring university status in 1992, a distinction is still being made between old and new universities. Local art schools have often been absorbed into polytechnics, while some of them were closed.
2   Sir William Coldstream was the chairman of the National Advisory Council on Art Education when the report was commissioned to outline the requirements for a new Diploma in Art and Design. He was an acclaimed professor at Slade School of Art (University College London).
3   Central Saint Martins is one of six colleges of University of the Arts, London. The student protest was present in social media through @occupyUAL on Twitter.
4   This event was heavily criticized as instrumentalization and commodification of artist's hardship (see conversations.e-flux.com/t/frieze-art-fair-the-monetization-of-your-misery/2680).
5   Interventions listed in the report include planning protection, direct investment in under-occupied buildings and creative uses of city-owned properties.
6   I graduated with a Fine Art degree from the University of the Arts, London. As part of this, I gained insider knowledge of the London art world as well as access to the networks of art school lecturers and graduates who are engaged with the alternative art schools.
7   See London Borough of Hackney 'Proposed Site Allocations Local Plan: Post Submission Modifications Version April.'
8   CREATE explores the relationship of artists with the city, and acts as agency to forge links in this field (http://createlondon.org/).
9   The expanded field discusses how art has crossed material and conceptual boundaries away from its original material specificity, with relational art being one example that sees art in the context of the everyday and communities as subject-matter or platform for artistic practice.

References

Artquest (2014) 'Open School East Reader'. [online] Available at www.artquest.org.uk/wp-content/uploads/OSE-Reader.pdf [Accessed 15 Aug. 2016].

Artquest (2015) *Artquest / Current Projects / System Failure*. [online] Available at www.artquest.org.uk/articles/view/system_failure [Accessed 10 Jan. 2016].

Asquith, S. (2014) 'Let's change the world for art students in 2014'. [online] *The Guardian*. Available at www.theguardian.com/education/2014/jan/06/art-students-2014-campaign-change [Accessed 10 Jan. 2016].

Banks, M. and Oakley, K. (2016) 'The dance goes on forever? Art schools, class and UK higher education'. *International Journal of Cultural Policy*, 22 (1): 41–57.

Bourdieu, P. (2011) 'The forms of capital (1986)'. *Cultural Theory: An Anthology*, 81–93.

Bourriaud, N. (1998) *Relational Aesthetics*, Dijon, France: Les Presse Du Reel.

Butler, T. and Lees, L. (2006) 'Super-gentrification in Barnsbury, London: Globalization and gentrifying global elites at the neighbourhood level'. *Transactions of the Institute of British Geographers*, 31 (4): 467–487.

Chatterton, P. (2000) 'The cultural role of universities in the community: Revisiting the university-community debate'. *Environment and Planning A*, 32 (1): 165–182.

Chicago, J. (2014) *Institutional Time: A Critique of Studio Art Education*, New York: Monacelli Press.

Comunian, R., Faggian, A. and Li, Q. C. (2010) 'Unrewarded careers in the creative class: The strange case of bohemian graduates'. *Papers in Regional Science*, 89 (2): 389–410.

GLA (2014) *Artists' Workspace Study*. London: Greater London Authority.

Hamnett, C. (2009) 'Spatially displaced demand and the changing geography of house prices in London, 1995–2006'. *Housing Studies*, 24 (3): 301–320.

Heygatewashome.org (2016) *Heygate was Home: Broken Promises* [online] heygate-washome.org/displacement.html [Accessed 14 July 2016].

Ivison, T. and Vandeputte, T. (eds.) (2013) *Contestations: Learning from Critical Experiments in Education*, London: Bedford Press.

Krauss, R. (1979) 'Sculpture in the expanded field'. *October*, vol. 8, 31–44.

Lees, L. (2014) 'The urban injustices of new Labour's "New Urban Renewal": The case of the Aylesbury Estate in London'. *Antipode*, 46 (4): 921–947.

Londonischanging.org (2016) *London is Changing* [online] www.londonischanging.org/ [Accessed 17 August 2016].

Madoff, S. H. (2009) *Art School (Propositions for the 21st Century)*, Cambridge, MA: MIT Press.

Moore, R. (2015) 'London: the city that ate itself'. [online] www.theguardian.com/uk-news/2015/jun/28/london-the-city-that-ate-itself-rowan-moore [Accessed 17 August 2016].

Oakley, K., Sperry, B. and Pratt, A.C. (2008) *The Art of Innovation: How Fine Arts Graduates Contribute to Innovation*, London: NESTA.

Paris, C. (2013) 'The homes of the super-rich: Multiple residences, hyper-mobility and decoupling of prime residential housing in global cities'. In Hay, I. (ed.) *Geographies of the Super-Rich*, Cheltenham, UK: Edward Elgar, 94–109.

Pratt, A. C. (2009) 'Urban regeneration: From the arts 'feel good' factor to the cultural economy: A case study of Hoxton, London'. *Urban Studies*, 46 (5–6): 1041–1061.

Reades, J. (2014) 'Mapping changes in the affordability of London with open-source software and open data: 1997–2012'. *Regional Studies, Regional Science*, 1 (1): 336–338.

Tickner, L. (2008) *Hornsey 1968: The Art School Revolution*, London: Frances Lincoln.

Turps Banana (2016) 'Turps Banana / Art School'. [online] www.turpsbanana.com/art-school [Accessed 14 July 2016].

7   RENT Poet

Commodity, respectability, and scam in Los Angeles

Brian Sonia-Wallace

Los Angeles is a rising 'creative city' (Florida 2012): as a prominent LA food critic put it, 'The most Brooklyn thing you can do at this point is move to LA' (Riley 2016). The 2014 Otis Report on the 'Creative Economy' has some striking figures. one in seven jobs in the LA area is related to a creative field, with 40 per cent of California's total creative force living in this area (Kleinhenz *et al.* 2015). However, a massive creative economy does not mean equal access to participation in creative life, as the grassroots institutions supporting the arts founder in a context of rising property values, and the artists relying on them compete for ever scarcer resources.

In this ethnographic chapter, I will attempt to systematize the theoretical framework revealed by my non-traditional practice of creating commercial poetry through busking and commission in Los Angeles. My practice started on the streets, crowd-funded by passers-by, but has quickly and unexpectedly put me into relationship with institutional art establishments, large and small. However, with an ethics, aesthetic and audience developed through direct contact with patrons outside of institutions, I find myself in the liminal space of contemporary art, as defined by Suhail Malik (2013): the anti-establishment becomes the thing most sought by the establishment.

Malik argues that art acts as an alibi for itself, hiding a network of power relationships involving arts institutions, brokers, buyers, sellers and the creative economy at large. I follow Malik in arguing that the most interesting thing in contemporary art, even for the artist, is not the art itself. It is historical structures and the artist and art's interactions with them. Under this model, through critical engagement, the artist has the potential to shift from being a muse, bringing pure contemplation, to a researcher generating important knowledge about society through active engagement in historical structures. Indeed, Houston and Pulido (2002) conceptualize the performativity of artistic practice not only as a space for social critique, but for the creation of social change. The central question I grapple with in this chapter, as a working artist practitioner chasing the almighty dollar, is: can art remain critical while simultaneously profiting off the system it critiques?

The following sections will explore three lines of thinking about art, activism and urbanism that continue to develop from and inform my progress in establishing myself as a career artist. The first section, 'Commodity', deconstructs my practice to unearth questions about art versus commerce and criticality versus mainstream accessibility. The second, 'Respectability', digs into the critical versus mainstream dichotomy to examine the changing landscape of platforms for the public to experience the arts and the effects of this on the art itself. The final section, 'Scam', presents capital as a technology in itself and examines how the arts may harness and domesticate this technology rather than submitting to it and being subsumed by it.

Commodity

I arrived at my commercial poetry practice almost coincidentally, with a background not in poetry, but activist theatre, which has informed my approach and ethics. I ran a theatre company in the United Kingdom as a student, from 2007 to 2011, called Tabula Rasa ('blank slate') Productions, with the explicit goal of creating devised or collaborative theatre with student actors. In devised work, the script originates not from a writer or writers, but from collaborative, usually improvisatory, work by a group (usually, but not necessarily, the performers) (Milling and Heddon 2005). In this process, my role was to bring a concept, devise games to explore it, and to record and edit the resulting text that the cast created. Through this curatorial role, I grew increasingly interested in social practice art, culminating in a dissertation on 'Augusto Boal and Theatre of the Oppressed', which shifts the lead role in creative projects from an auteur creating *for* a community to a provocateur leading a community through its own self-realizing creative process (Sonia-Wallace 2011). As an art form, theatre is generally more public and collaborative than writing, and prizes spontaneity and interaction – skills that would become essential as I began to develop my unusual practice of creating poetry collaboratively with strangers for cash.

In 2012, when I moved back to my native Los Angeles, I heard a story on National Public Radio about a poet in San Francisco who made enough money to quit his day job, just writing poems for passers-by in Golden Gate Park on a manual typewriter (NPR Morning Edition 2012). I was inspired to borrow a typewriter and try this out at a few events, and found that I actually averaged $30–50 an hour over the two to three hours of an event, well over the $9.50 per hour state minimum wage, meaning that just a few hours of work at the right venue could surpass the pay for a full day of unskilled labour. In September 2014, born of economic necessity and inspired by my involvement with the movement to legalize street vending in Los Angeles, I challenged myself to pay my entire rent through busking poetry on the street and at events – I called it the 'RENT Poet challenge'. That first month, I ended up making minimum wage, and accidentally birthed my ongoing performance art project/experiment/business: RENT Poet.

I view my RENT Poet practice not as a new endeavour, but a continuation of my participatory and voicing theatre projects, as all the poetry I write is collaborative: it is based on the stories and topics people give me.

RENT Poet is a vision of poetry as a service industry, rather than 'art for art's sake'. People are paying me to write poems that relate personally to their lives, and the academic quality or artistic integrity of the poetry is secondary to how it makes the people standing in front of me feel. The process of busking poetry has numerous theoretical aims: demystifying the writing process, making poetry personal and relevant to individuals, and modelling a society where anyone can be a patron of the arts with the responsibility of commissioning artistic work. Rejecting the notion of art for art's sake in favour of Malik's claim that all art is political, my busking practice began as a progressive project on the contested but largely accessible forum of the street. My highest earning public events, which I began seeking out, were street fairs, which in Los Angeles often took the form of food truck festivals, art walks and community events. These events were free and accessible, but came with a sizable middle-class attendance, a focus on experiential value, and an expectation that patrons would encounter food and other items for purchase at about my average tip range: $5 to $10 a poem.

This notion of value in busking or tip-based work is co-created by the artist, telling the public their work is worth money, and the public willingly paying for it or walking away. My typewriter case reads, 'Presence is a Commodity', something a patron told me at one point after ordering a poem and waiting, with perfect attention, for ten minutes as I wrote for her. In a broader sense, RENT Poet confronts the way in which artistic practice is contingent not on any real or tangible value, but on a subjective and speculative value. The presence of both artist and audience is inextricably linked to the product here, drawing conceptually from work like Marina Abromovic's 'The Artist is Present' (Arboleda 2010) and from Marcel Duchamp's notion of the audience completing the artwork (Duchamp 1957). The only difference between sitting on the street corner begging and sitting in the same place writing poetry is an audience buying into the idea that the product here is art, and, thus, work with value. The perceived value of the work in fact serves to legitimize my mere presence on streets that have not come a long way since *City of Quartz* (Davis 1990), on which the homeless and immigrant food vendors are frequently ticketed and moved along.

Based on the success of my first month busking, institutions began to approach me about making art within their contexts, creating an ongoing moment of crisis in my work by changing the relationship between audience, poet and poem. Before, with RENT Poet, I would ask members of the public to evaluate the worth of my labour, independent of institutions that would typically guide them. In doing so, it shone a light on how we place value on art, and especially the artist's presence, in the contemporary world, poking fun at the absurdity of twenty-first-century art and labour commodification, while simultaneously relying on this absurdity to sustain itself. The catch-22 I encountered was what it meant to translate the critical framework and grassroots origins of my practice to the formal playground of institutional art, which I had sought to explicitly to critique. But, as an artist frustrated by lack of resources, I wanted institutional support, seeing it as a route to a platform, an audience (sometimes), and a sense of legitimacy for myself.

It is worth noting here that a minimum wage salary in Los Angeles is not a living wage, and increasing institutional support for artists becomes an aspirational endeavour toward a lifestyle above the federal poverty line. This puts more emphasis on the tensions artists face between 'selling out' (being able to afford a modest apartment) and 'sticking it to the man forever' (living in poverty or through the generosity of others). Here, refusing institutional support may therefore be a marker of privilege as much as an ideological platform – only those with the resources to make art for art's sake may escape institutional demands. It raises the question: can simple pragmatism be divorced from ideology or a personal manifesto?

For the commercial gigs, booking agents would misremember my name as 'Rent-a-Poet', tone-deaf to the connection with 'rent boy' as a slur that exists in the United Kingdom and gay communities. As a cultural researcher and artist, words and especially names hold power. The name RENT Poet both lays out my goal (paying rent) and sensationally alludes to a 'rent boy', or male prostitute. My initial goal with the name was to elicit the same arguments that people have used to condemn sex work for generations – that surely something as personal and intimate as poetry should be reserved for real intimacy and deep-seated personal feeling. It seemed to me that there was a tension inherent in using poetry for a utilitarian, maybe even counter-revolutionary goal like financial profit, no matter how inclusive my practice was. By making this provocation explicit in the name itself, I sought to establish the RENT Poet project as punk and iconoclastic, a scheme to expose the underlying rationale and function of artistic work – not high art, but a political practice driven by financial forces and cracked open, by surprise, to the general public. But my work, intended to satirize commercial art, and, indeed, the notion of labour capitalism, had itself, inadvertently, become commercial.

In the fray, I somehow wound up featured in *The Knot*, a national high-end wedding magazine, the pinnacle of cultural consumption. It was a joke, I swear!

But the irony and real biting power of this joke is how it reflects a moment of post-modern criticism in art where satire looks like Stephen Colbert on *The Colbert Report* – it becomes the thing it comments on in order to mock it (McGrath 2012). This creates a catch-22 of identity: individuals unite behind something they don't believe in with a knowing wink that they are in on the joke, while others take the statements of the trickster at face value and unite behind them because they *do* believe in them.

Alongside the name, my performance and writing practice simultaneously decry and embrace their place within a commodified world through the irony of what I call 'capitalist poetry' – that is, intimate poetry written with capital exchange value as its first goal and art as its second. Examples range from writing love letters for investment bankers to their (absent) wives at a conference whose only female attendees were paid models, to writing a 'multi-cultural, non-denominational, Holidays-in-Los-Angeles' poem for county marketing materials – only four, unconnected lines of this poem ended up on the flyers. A layman's definition of commodification is the transformation of the essential or un-ownable (e.g. love or poetry) into a product to be bought and sold (Merriam Webster 2016). Under this schema, my goal was to commodify poetry. I view capitalist poetry itself as a piece of performance art that highlights the absurdity of capitalism in being able to fetishize and commodify anything in a service economy, even poetry. But in a deeper sense, my project is about contesting the definition of commodification and how art is valued.

A Marxist analysis adds nuance to the layman's definition of commodification used earlier through Marx's conception of alienation. Marx defined commodity as something where the creator is estranged from the object created, hence losing the connection to the labour required to create it. In my busking practice, I type in front of the purchaser, creating a visible link between the creator, the creation and the person enjoying the fruits of my labour. Arguably, the value I create comes from decommodifying and communalizing the act of writing, which is traditionally completely hidden from the reader and performed alone. The use of a manual typewriter is itself symbolic of the 'labour' involved, not just a curiosity but a way to spotlight, quite literally, the means of production, which in this practice become directly and intimately available for hire. Part of what RENT Poet can provide is a good Marxist model of artistic labour without alienation – a collaborative art project that deconstructs specialist hierarchies in what may be a fitting theme for an artwork. To this end, I write a lot of poems about people's pets, for example.

The institutionalization of my work through pre-commissioned appearances closes off my work and stratifies it according to the audience of the company, individual or artistic institution that commissions me. As I engage in more of these 'rent-a-poet' appearances, seeking financial stability, the direct relation with my audience as micro-patrons, like micro-investors in an enterprise, begins to fade. Patrons' interactions with me become risk free for them, reducing the charge of the experience and relegating it to the status of curiosity. At live events, I become merely part of the entertainment landscape, working alongside magicians, caricature artists, psychics and the like. I collected a number of my best street poems and published them at the beginning of 2016 as a book called, *I Sold These Poems, Now I Want Them Back.* The tongue-in-cheek name, like RENT Poet itself, satirizes the possibility of re-creating the personal and intimate experience of work in which the artist is present for a general readership.

Respectability

My initial challenge was to pay my rent using poetry, and from the beginning, I knew street busking would only get me so far. In doing my work, I perceive a looming dichotomy between guerrilla-style art created at the fringes, usually for free as passion projects, and art in the eye of the 'polite society' of older, wealthier patrons – art that receives institutional support, resources and longevity (in the form of documentation, reviews and analysis). I define this difference as 'respectability'.

Since starting RENT Poet, I have been trying to sneak my typewriter into polite society, seeking out the big cultural institutions, corporate gigs and weddings. The desire to publish my book was born of the realization that having formal literary credentials might increase my desirability for events. However, the twenty-first century has changed the relationship between artists, audiences and institutions in ways that continually make me call into question my original presuppositions about the value of this artistic respectability.

Because I was an emerging artist busking on the streets, the audience funding my work was itself made of workers outside of institutional art, disproportionately diverse in age, ethnicity and income for patronage's usual suspects in Los Angeles. I was able to find an audience by going around institutional relationships and measures of quality. As neoliberalism has retrenched and downsized many state-funded institutions, a surge of crowd funding, grassroots projects and digital pathways is being born out of (and not despite of) neoliberal processes. The environment created is certainly more competitive, if more open, accessible and participatory.

My work is mirrored by, and takes advantage of, advances in technology which change the models of how work is created and sustained. Because technology enables artists to reach massive audiences on their own, it cracks open the door to bypass the cultural institutions that serve as intermediaries between creators and consumers. Social media, like Instagram, Twitter and Facebook, allows independent artists to cultivate audiences of friends or shared communities of interest organized around hashtag themes. From Justin Bieber to comedians, YouTube and Vine artists are surpassing traditional artists in scope and critical/commercial success.

Crowd funding, like Kickstarter or my own street busking practice, allows artists to get funding directly from these enhanced audiences. The top poet on Instagram, who self-published a book on Amazon, outsold last year's Pulitzer Prize winner in poetry by a factor of 100. Of course, institutions that mediate artistic quality are scared – though @rmdrake's poetry may be of questionable depth, millions of people read it. Technology also allows for greater scales of unmediated interaction between artists and consumers and customization of art. This new world changes the role of institutions within the arts, taking away much of their status as gateways to audiences and funding, and relegating them to a support role: megaphones and tokens of legitimacy that enable these essentials to grow. In this new landscape, the value of institutions to artists becomes contested, though it may be argued that this simply represents a shifting elitism, as not everyone has access to these new pathways to success. Additionally, artists and audiences relying on these pathways can lose the valuable potential for institutions to play the roles of mediators and, sometimes, active instigators and incubators of creative quality.

The emerging grassroots (or 'start up'?) creative economy, fuelled by social media and crowd funding, still operates within a context of capital, but places the onus of artistic support on local communities rather than large institutions as artists turn to their communities for funding and support (or the market has placed the onus). One of the pitfalls I recognized in my artistic practice was that my liberation from institutional 'work' to pursue art as RENT Poet was contingent on the continued drudgery of my grassroots supporters and friends. When I first started RENT Poet, I would go and write anywhere, making as little as $7 in a day. This model also barely made a subsistence level of income, and I realized the need to institutionalize to make a living, which meant prioritizing making more money in less time: writing personal work for fewer people. In practice, this meant institutional relationships with everyone from private parties to corporate events to museums, publishers and theatres – moving from RENT Poet to Rent-a-Poet.

As part of this shift, I have been involved with organizing other typewriter poets to form the Melrose Poetry Bureau, which has traditional representation and seeks only gigs large enough to afford multiple poets – often with large tech and entertainment companies outside of the traditional poetry world. Groups like Haiku Bros in New York and Typewriter Rodeo in Texas follow similar models, while individual poets like Jaqueline Suskin pursue high-value, single-poet gigs.

This model of gig-to-gig artist work, which falls short of progressive labour movement desires for stability and security for workers, is nothing new for artists, but increasingly looks like the rest of the working world in the twenty-first century. The insecurity of artistic employment seems less daunting in a reality where companies seek to contract labour-on-demand as consultants, rather than full-time employees, a world where Uber is replacing taxis and Airbnb is replacing hotels. Creative destruction – Schumpeter's idea – does this also apply to the art world?

In moving away from crowd-funded art and toward institutionally-funded art, I am moving from a paradigm of art's value as labour value (an hourly wage or per-poem wage) to art as speculation (an appearance fee), which derives its value not just from services rendered but from the speculative future value of the artist. The very presence of the artist moves from Marx's conception of use value to exchange value. But this shift may be one devoutly to be wished, as artists recruit institutions to invest in them and their success. True to Malik's hypothesis, I have found that, despite tracing my origins to street theatre and Marxist art, my critical art is absorbed and *wanted by* the very institutions it critiques. Malik understands this as an institutional quest for the 'real' aka 'non-institutional' art (Malik 2013). The more ironic, critical or abrasive the artist, the more value they create for institutions profiting from their work through speculation in the success of their art. The street artist Banksy is the poster boy for Malik's critique, a graffiti artist who started decorating walls illegally, risking arrest and fines, and has turned a movement that was condemned as causing urban blight into an indicator of wealth and status. This creates another catch-22 – by railing against or bypassing a system, I have created the tools for my own access to that very system. Even with the present chapter for this volume, I was invited to contribute writing (to institutionalize my thoughts) after a Facebook rant about institutional access and abuse of artists.

This tension between critical projects and institutions is well documented across movements. Sociologist bell hooks (Gloria Jean Watkins, lower case spelling of pen name intentional) speaks of how dominant culture 'eats the other' in absorbing and reselling the revolutionary or post-modern (hooks, 1992). Like Malik, she shows how the anti-establishment becomes the thing most sought by the establishment in what she describes as a hunger for the 'primitive'. hooks writes about the commodification of black nationalism as symbols of revolution become co-opted as fashion statements that are less about collective action and more about personal identity. Mariana Torgovnick expands these views from an anthropological/post-colonial framework, explains this 'fascination with the primitive' as a product of a cultural crisis in identity in the West, where the desire for alternative ways of experiencing the universe collide with the need to clearly demarcate subject and object (Torgovnick 1990).

This desire for experience is the underlying need that my practice meets for the public. I often think that much of what I do with RENT Poet is an exercise in vicarious living – beyond the poems themselves, I am selling good white-collar employees a romantic notion of the artist, up close and personal by co-creating work with them. By brazenly taking on the idea of 'starving artists' through using art in direct dialogue with the public to pay rent, I give those I write for the vicarious experience of quitting their jobs and walking away to pursue a 'higher calling', without actually having to walk away (though I have become something of a job quitting consultant in my circles). Torgovnick critiques cultural absorption into the mainstream because the demarcation between subject and object creates a safety in encounters with the 'other' by situating it neatly into a dominant framework for understanding the universe. Actual change in thought, based on these encounters, becomes impossible, because the ontology that sees expressions of culture uses the eyes of a hegemonic culture that allows these expressions only in its visible spectrum and according to its own models of understanding. For instance, my practice is seen by those who seek me out with their job woes as 'doing what I love' rather than as a form of low-income entrepreneurship – another type of work. And, of course, really it is both, an intersectional practice with multiple and overlapping identities, aims and intentions.

Alongside the ontological critique, an activist art critique would hold that, in showing a model of artistry-as-livelihood, while not building a sustainable path for others to achieve it, I am actually completing a counter-revolutionary project of building a fantasy to keep people happy and sedated in their lives of drudge labour. The irony is, of course, that creating and sustaining the fiction of the wild and free artist is a lot of drudgery and work. I am working on this – my practice increasingly focuses on finding ways for artists and the art community at large to sustain themselves, but these actions feel like finding loopholes in a broken system, rather than systemic change.

I acknowledge the privilege inherent in being able to do what I do: I am a young white man, I come from a supportive middle-class family, I got a degree without debt, and I worked at a nice white-collar job and saved up money for a couple years before embarking on this path. I recognize the need to continue evolving as an artist, which is why I'm trying to parlay my tenuous financial stability into entrenchment in the cultural institutions that give credibility and create livings for 'career artists'. I had a head start, and I'm still playing catch-up in a world where the artist is a cultural entrepreneur, subject to the same risks, warnings, and start-up needs as a small business owner.

The systemic or institutional access to resources that I have achieved, still in its infancy, has raised its own set of issues. I believe that this first layer of inclusion represents just a shift from 'marginalized from without' to 'marginalized from within'. Since returning to Los Angeles, I have worked with small theatres and continue to find their model disturbing and exploitative of artists. Most theatres co-produce events with artists, splitting a portion of the money made. However, the theatres are guaranteed a certain baseline each night while the responsibility of generating an audience falls almost entirely on the artist, not the theatre. This means that, as with music at a small scale, in many cases artists bring audiences, ticket sales and their work, and only the venue makes any money. These institutions are set up to assume the minimum amount of financial risk, passing it to the artist, instead. One can hardly blame them – they are, almost universally, renters, subject to the whims of landlords whose primary concern is market value. In some of these institutions, it is possible for artists to make a small amount of money so long as the artist fills the theatre every night – in most cases, this money is negligible compared to the quantity of work involved.

I have heard similar criticism levelled at the theatre festivals, such as the Edinburgh Fringe Festival, which are theoretically platforms for new voices to cheaply produce work and gain exposure to build future career success. Critics argue that the festivals exist to profit venues on the back of free artistic labour that not only produces art, but also promotes it (Slayer 2012). Seeing the new technological capabilities of artists for reaching and growing audiences organically, venues (and through them, property owners) are cashing in. Even the new grassroots access technology opens to artists in reaching audiences is funnelled to enrich entrenched power in institutions. The line, it seems, is blurry between providing a platform for new voices and exploiting labour, in this case.

In the context of Los Angeles, this conflict between the vision of venues as providing an artistic home, versus as exploiting artistic labour, came to a head in 2014 with the start of the ongoing '99-Seat Theatre Wars' in small theatre (Miller 2015). In 1988, Actor's Equity agreed to waive payment for actors working in small theatre houses with limited performances, acknowledging artists' desire to create in a city with an overabundance of actors for the limited resources available. Under these rules, actors received $7 per performance (not per hour), with no rehearsal pay, while administrators made salaries as non-profit employees. However, in 2015, despite the support of majority of membership to keep this plan in an advisory vote, Equity announced that it would update the 99-seat rules to comply with labour law, requiring theatres to pay minimum wage for hours worked (McNulty 2015). This has created the ironic situation of union members suing their union to remove protections (the lawsuit is ongoing), fearing that paying actors will cause venues to close and less art to be produced. Part of my move from theatre to poetry was a desire to distance myself from this battle and find alternative modes of creating work.

Wrapped up in the issue of access to the profits of labour is the question of spatiality of art and capital. Does it behove artists to find an artistic home where audiences can expect them and grow that base, at the cost of most of their profits, or to be nomads following flows of capital between spaces and events, but without the potential for stable growth that institutions provide? As a poet, I can maximize my profits by appearing for short, high-paying gigs at weddings and corporate events, but this also comes at the sacrifice the perception of 'respectability' that surrounds working with traditional cultural institutions. Not only this, but a regular artistic home provides an incubator for new work and can help in the cultivation of an audience who will be more adventurous, knowing the artist, to go outside their comfort zones and take a chance on new, less immediately accessible work. Mid-level theatres and artistic companies often use a mixed model as a compromise, creating work at a home base at a loss, financed by donations, and choosing certain work to commodify and elevate to a financial product that then tours other venues. Even for these relatively successful companies, a look behind the scenes shows a frenetic reality in which the income generated isn't enough to support the volume of work that goes into sustaining this model, resulting in high burn-out and little room for experimentation and growth.

Looking at traditional models of financing start-ups in the current economic system, it is no wonder that individual artists and small companies have a spectacular failure rate. A tendency I have noticed among artist friends in Los Angeles is that, while droves of artists migrate here, after two years, like clockwork, most move back home and/or abandon artistic aspirations altogether in favour of stable careers. As boutique entrepreneurial enterprises, artists exist in a market where, to succeed, art must meet the criteria of commodity (capable of making money) and respectability (capable of having audiences who bring that money). In Los Angeles, most independent artists I meet subsidize their artistic careers, which mostly make the individuals a loss at a tiny profit to wealthy landlords. These artists subsidize their artistic practice by earning low wages from day jobs as bartenders, servers, personal trainers, teachers and other flexible service work. Why these types of work are valued more highly than art is unclear, except perhaps that their service function is more explicit. In an expensive city like Los Angeles, the wages earned from these 'day jobs' are sometimes enough to cover living expenses, but rarely enough to invest in an artistic brand or product so that it grows like a traditional business and attracts clients. Scam rears its ugly head here, too, in the form of the artistic pyramid scheme: trying to invest in themselves as a product, artists face a dizzying array of classes taught by other artists at all levels of career and quality, producing a culture where artists live off one another by teaching each other classes. Sometimes a whole course of expensive classes is required to even be considered for artistic companies, and the majority of classes I have seen centre around the craft rather than the business of art.

California, and the United States in general, set aside only a small percentage of their large budgets for arts funding, a civic footnote. To access the majority of funding, foundations and government entities require that artists institutionalize themselves into tax-exempt 501(c)3 non-profit organizations. Non-profit status carries huge burdens of administrative work and overhead, necessitating that artists spend a majority of the little funding available to hire and maintain a large, non-artistic support staff. It is my view that LA's 99-Seat Theatre Plan is an evolution of this model – paid administrators worked exclusively with volunteer actors, for dozens of years, with the blessing of the government and unions, designed to protect artists. Even within the structure of a non-profit, funding is usually restricted to specific projects (products) and populations (clients), explicitly restricting the growth of arts as a business by placing the limitations on it that we place on charity.

With funding limitations in place, small arts institutions (many of which I believe are just individual artists with support staff) are unable to grow without passing the financial burden onto other artists, which means that the primary function of the artist must be not to create art, but to bring new audiences and, with them, revenue streams. The artists' value is contingent not on their artistic ability, but on their ability to create free publicity for their work and, as a by-product, their venue. Because part of my work has consisted of developing audiences outside of traditional venues through street theatre, busking, side-show acts and participatory workshops, I have gained a perceived value to venues that is largely contingent on bringing these audiences to their spaces.

When it comes to the artwork itself, there is a cost-benefit equation in which artists' need for platforms to reach audiences is balanced against the unique needs of each platform for work suited to it. People upload content for free to YouTube or Instagram or Facebook to reach a wider audience, faster. Artists can use cultural institutions with a similar intent, though the barriers to entry and investment costs are higher. Because of this, that cost of entry (financial and in terms of compromises in the form of the art) must be balanced in each case against the benefit to be gained from the respectability or prestige brought by institutions. Since smaller institutions do not bring their own public, artists often work with them not for their audiences, but in the hope that the appearance of institutional prestige will lead to working with progressively larger institutions who have more resources and access to their own audiences. It becomes a relationship between the medium and the message – why should my brand create content for your platform? How can I know which platforms are worthwhile?

The artistic work itself begins to reflect the reality of artists-as-nomads, bringing their own tribes, and stopping at oasis venues or riding with a product for a limited time. In street theatre contexts, my work does not exist independent of the audience but as a conversation with them; however, to fit into institutional spaces (stages and art galleries), I have to make the work itself more presentational and less intimate and participatory to suit the format. This changes my own relationship to the audience in creating work, and what kind of work I create. I began with trying to break artistic hierarchies and blur the lines of what it means to consume versus produce (support versus generate) art, but find my practice has begun to reinforce these very hierarchies, the distinctions between creators and consumers. By bringing my work into their spaces, technology and institutions alike change the work itself – the next section 'Scam', will return to the philosophical underpinnings of RENT Poet to evaluate that change and propose a solution that reconciles high art and commerce.

Scam

Art!
For people who want to be tortured!
Experience alienation, resentment, loss, and rage
from the comfort of your seat.
Enjoy the ride.

—RENT Poet

A scam is an attempt to defraud a person or group by gaining their confidence. In the context of this chapter, scam overarchingly refers to entrenched systems of unequal access or injustice for artists – the way capitalism appropriates art and the way institutions exploit artists. But the various scams I have witnessed may not be an antithesis to my artistic production. In another light, they may be viewed as an enabler of that production. In this chapter, I will have my cake and eat it too – I will argue for profiting from a dystopic system by using it to model utopia. To do so, I must convince you that the dystopia I present is actually a mechanism to enable utopia. This chapter is titled 'Scam' for the same reason my practice is called RENT Poet and my book is called *I Sold These Poems, Now I Want Them Back* – I am warning you, up-front, that the very essence of this chapter may well be a scam – an attempt to reconcile commodification and institutionalization with a sense of optimism for art. I am perpetuating this scam on you, the reader, and I am doing it with a grin, a big snake-oil salesman grin, and with the confidence that you will still buy it.

RENT Poet:
(in a Carnival barker voice) Right this way!

New contexts of technology and work affect not only the production and consumption of art, but, more interestingly, I argue, shift our capacity for emotion and humanity – the themes and power of the art itself. I'm fascinated by how the content of art, specifically poetry, will change in the face of new technology, not just how its distribution will change. I've started writing poetry in emoji, pictorial characters, and am very excited at Facebook's move to include automatic emoji as responses. Writing is getting closer and closer to speech – it is all instantaneous, and now we are bringing in pictograms to show tone of voice. More people are writing in the twenty-first century than have ever written before in the course of human history – even if it's only Tweets – which I see as a profoundly democratizing force – and language is changing with this technological revolution in access to writing and communication. Labour, also, becomes mediated by this technology, as anthropologist David Graeber writes about a move away from production and toward what he calls 'bullshit jobs', an information economy based on exchanges of knowledge and power where resources are moved around but never produced (Graeber 2013).

My own practice exists in this context of flowering access to the written word and the changing face of labour. Defining my work as below-poverty subsistence living, or as boutique entrepreneurship, does little to encompass the lived experience of the work: I wake up and do what I want. It's the punk slacker dream with a veneer of business sense and a good gimmick. One of my explicit goals is not to work that hard, denouncing a puritan work ethic that glorifies work for its own sake as much as I denounce an ethic that glorifies art for its own sake and derides its political and commercial implications. My goal is not to make the most money possible, it's to *need* to work as *little* as possible and to do the most fulfilling work possible in that time, leaving the rest of my time open for bettering society or watching cats on YouTube – a rather transcendentalist notion of personal freedom.

And so, technology becomes democracy, and laziness, personal liberty. The irony is thick.

This ironic foundation of RENT Poet locates it squarely within the post-modern framework of awareness of contradiction, playing art for art's sake against commodification, authenticity against respectability. RENT Poet against Rent-a-Poet. This framework of post-modernism may be seen as paralytic rather than revolutionary – seeing and contemplating helplessness (creating 'pure art' in the face of market forces) in such a way that re-enforces and creates helplessness rather than fighting it. However, I take an accelerationist view of art's progress in the face of ironic helplessness and artificial value. I want to believe the accelerationist prediction that technological advances will lead to plenty of resources for all, without anyone having to work (Williams and Srnicek 2013) – with all the time and no duties, the argument goes, everyone becomes philosophers and artists. Plato grins from his grave. I see this as preferable to the alternative future I see manifesting itself, where everyone works in human resources and marketing. As a poet, I often feel like a fraud, like I am perpetuating the scam of my art on the world. But in a context of post-industrial work, a world of on-demands labour, information economies and speculative value, my art suddenly feels less like a scam and more like a natural part of the landscape. RENT Poet is a bullshit job, but so are the vast majority of contemporary jobs.

Many artists I talk to hold as a utopian ideal, a world where art is possible without money – a world of depoliticized art. Their call smacks of a philosophical 'return to nature', to an egalitarian society where art exists only 'for art's sake', but with the material conditions created by the post-industrial capitalist system that make all aspects of life a commodity. Accelerationist ideals, in contrast, hold that capitalism and technology can, themselves, become revolutionary forces to enable a new synthesis from this contradiction and that movement towards this utopia should this be accelerated, not retarded. The absurdity and contradictions of capitalism are not an enemy to fight against, but a doorway to pass through.

Rather than a Marxist push for human resources and marketing to return to manual production or farming, occupations which create real value by creating new resources, accelerationists argue that we should embrace the non-productive nature of their work. In fact, we should push for them to further abstract their labour from the creation of value! When this happens, don't these professions become, essentially, artists? With regards to labour in particular, accelerationist views hold that the 'end of work' is not only possible but imminent: that mechanization and prosperity are replacing human labour with machinery and in doing so are breaking the relationship between capital and labour. With the need for sustaining resources divorced from the need for work, people gain personal freedom – the freedom to shift their focus from wage labour, built on an old model, to pursue more meaningful tasks of self-actualization and societal improvement. Where this model falls down in practice is in the distribution of the spoils of mechanized production, which tend to cluster around the few who own the machinery simply because of their ownership, and leave out the rest of society. Accelerationists push for enhanced technological progress and a social society that distributes the fruits of its labour. Experiments are already underway with mandatory minimum incomes in parts of Europe, a basic living stipend for all residents based on the notion that, in industrialized countries with more than enough resources for all citizens, human suffering for want of resources is a human rights abuse. In the industrialized world, we already live in a context with plentiful resources – it is mainly our distribution systems that are broken.

In a sense, the issue of non-essential production has always been at the heart of artmaking, where artists do not create essential resources and so rely precariously on wealthy patrons or sponsors, or the state, to support them. But the idea of 'art for art's sake' becomes redundant in a world without work, where everything is for its own sake. The activism of my practice, if any, is not a project of revolution for a specific group, but a small model of what this utopia might look like. What I sell explicitly is poems, but my underlying product, or brand, is vicarious living in a fictional twenty-first-century America in which people can sustain themselves without traditional labour, and so have time to pursue their passions. This product, of course, is a scam, a fiction, because in the current system I am actually working like a dog. But the service I provide is as real as most contemporary work – a science fiction narrative with a carefree artist as its protagonist, exploring the mechanics of an accelerationist universe. Through this imaginary exploration, I am taking part in the real project of building this new world. Please don't tell my corporate partners.

This is my scam.

References

Arboleda, Y. (2010) 'Bringing Marina Flowers'. *The Huffington Post.* [online] Available at www.huffingtonpost.com/yazmany-arboleda/bringing-marina-flowers_b_592597.html [Accessed 7/31/2016].

Davis, M. (1990) *City of Quartz.* New York City: Verso Books.

Duchamp, M. (1957) *Session on the Creative Act.* Houston, Texas: Convention of the American Federation of Arts.

Florida, R. (2012) *The Rise of the Creative Class, Revisited.* New York: Basic Books.

Graeber, D. (2013) 'On the Phenomenon of Bullshit Jobs'. *Strike Magazine.* [online] Available at www.strikemag.org/bullshit-jobs/ [Accessed 7/31/2016].

hooks, b. (1992) *Black Looks: Race and Representation.* Boston, MA: South End Press.

Houston, D. and Pulido, L. (2002) 'The work of performativity: staging social justice at the University of Southern California.' *Environment and Planning D: Society and Space* 20 (4) 401–424.

Kleinhenz, R. A. *et al.* (2015) *Otis Report on the Creative Economy.* [online] Available at www.otis.edu/sites/default/files/2015_Otis_Report_on_the_Creative_Economy_CA .pdf [Accessed 7/31/2016].

McGrath, C. (2012) 'How Many Stephen Colberts Are There?' *New York Times.* [online] Available at www.nytimes.com/2012/01/08/magazine/stephen-colbert.html? pagewanted=1&_r=1& [Accessed 7/31/2016].

McNulty, C. (2015) 'In 99-Seat Tussle, Don't Lose Sight: Common Goal Is a Healthy Theater Scene'. *Los Angeles Times.* [online] Available at www.latimes.com/

entertainment/arts/la-et-cm-ca-mcnulty-99-seat-theater-notebook-20150423-column. html [Accessed 7/31/2016].

Malik, S. (2013) *On the Necessity of Art's Exit from Contemporary Art*. [online] Available at   www.artistsspace.org/programs/on-the-necessity-of-arts-exit-from-contemporary-art [Accessed 7/31/2016].

Merriam Webster (2016) *Definition of Commodify*. [online] Available at www.merriam-webster.com/dictionary/commodify [Accessed 7/31/2016].

Miller, S. (2015) 'Equity Unveils 99-Seat Plan calling for L.A. Actors' "Minimum Wage"'. *American Theatre Magazine*. [online] Available at www.americantheatre.org/2015/02/20/ equity-vs-l-a-99-seat-theatre-the-final-showdown/ [Accessed 7/31/2016].

Milling, J. and Heddon, D. (2005). *Devising Performance*. Basingstoke, UK: Palgrave Macmillan.

NPR Morning Edition (2012) 'A Poem Store Open for Business, in the Open Air'. [online] Available   at   www.npr.org/2012/04/17/150722541/the-poem-store-open-for-business [Accessed 7/31/2016].

Riley, D. (2016) 'Inside the Mind (and Mouth) of the Country's Gutsiest Food Critic'. *GQ Magazine*. [online] Available at www.gq.com/story/jonathan-gold-gutsiest-food-critic [Accessed 7/31/2016].

Slayer, B. (2012) 'The Edinburgh Fringe is the Real Pay-to-Play Scandal'. *Chortle Comedy News* [online] Available at www.chortle.co.uk/correspondents/2012/03/23/15107/the_ edinburgh_fringe_is_the_real_pay-to-play_scandal [Accessed 7/31/2016].

Sonia-Wallace, B. (2011) 'Toward a Green Theatre'. MA Dissertation in Sustainable Development. Scotland: University of St Andrews.

Torgovnick, M. (1990) *Gone Primitive*. Chicago: University of Chicago Press.

Williams, A. and Srnicek, N. (2013) *#ACCELERATE MANIFESTO for an Accelerationist Politics*. Critical Legal Thinking. [online] Available at www.criticallegalthinking. com/2013/05/14/accelerate-manifesto-for-an-accelerationist-politics/ [Accessed 7/31/2016].

8 Artistic activism as essential threshold from the 'peaceful, rational, non-violence' demonstrations towards revolution

Social actions in Hong Kong in the pre-umbrella movement era

Liza Kam Wing Man

This chapter analyses how artistic activism since 2007 has been a necessitated threshold guiding the aloof Hong Kong populace to start expressing their political quests, by subconsciously disguising them in artistic actions and performances. However, this chapter begins not in Hong Kong, but with a vignette relating to the work of a Hong Kong-based artist being staged in Weimar, Germany. I was a part of this site-specific performance, and will use this anecdote, and reflections on my experience, to set the stage for this chapter's discussion on art, activism, political meaning, and the context-dependent meaning attached to the urban space in which / through which this protest art is realized. Following a reflection on the performance in Weimar, I will move to Hong Kong, and discuss research I conducted there where I explored local residents' interpretation of meaning attached to historic urban spaces and their willingness – or not – to engage in art-based protest movements. This will draw upon Lefebvre's (1974) spatial trilogy and the sometimes contradictory relationship between 'perceived', 'representational' and 'lived' space. But first, we begin in the central square in Weimar, Germany, a place laden with (dark) historic meaning.

On one of the hottest days in summer 2015 in Weimar, Germany, the art performance, *One Sound of the Histories*, took place as part of the 'Kunstfest Weimar' (Weimar Art Festival) on Weimarplatz. Formerly known as the Adolf Hitler Square (Platz Adolf Hitler) during the Nazi period, the square is a monumental space, heavily charged with history. Surrounded by Nazi-style buildings that host the Weimar City Administrative Offices, the square is closed to the public on normal days. It was, after all, the square where Hitler reviewed the Nazi soldiers.

'One Sound of the Histories' was one of the first performances that was granted permission by the Weimar City Council to take place. Isaac Chong, a Berlin-based artist originally from Hong Kong, created and curated that day's performance. He invited more than sixty people to line up on the square to simultaneously tell their personal story/ history for fifteen minutes.

Departing from the concept of Chong's creation, as one of the 'performers' who participated in the storytelling and as a researcher going through Chong's documentations afterwards, I start this chapter with two points of discussions about this performance. The first one concerns Chong's process of acquiring permission from the Weimar City Council for the performance to finally take place and hence the perception of the performance in political terms implied by the necessity to go through such an 'application process'. The second point concerns the discrepancy between Chong's initial conception of his work and my interpretation, as well as the disenchantment induced through taking part in the performance as one of the performers. This initiates the discussion of how 'art' is perceived, what it is expected to do, what it can achieve in reality, and finally, how much/less the three could be different. From there, the chapter depicts the ideals articulated in the social movements in the pre-Umbrella Movement era in Hong Kong, with focus on the translation of the direct political activism to the earlier artistic actions, which aimed at provoking fellow citizens and the government. The diverse impact brought by the artistic actions are internalised by performers or audiences through their repetitive participation, which von Hantelmann referred to as 'performativity' (von Hantelmann, 2014). It is based on such foundations that the Hong Kong populace finally commence their reflection on the 'Peaceful, Rational and Non-violent' mode of protesting, through the slowly accustomed mode of contextualising ideals into actions in relation with the public space, and a frequent negotiation between the perceived and the lived (Lefebvre, 1974). Hongkongers are confronted with their local history and hence their identity.

Weimar: Art and its 'innocent' disguise

According to the documentations from Chong, while inviting people to take part in the performance, he was frequently questioned if the history and hence political meaning of the space had been taken into account during the conceptualisation stage. However much he had reflected on the political meaning and history of the space, in his application for permission to conduct the performance, he had to articulate cautiously and illustrate the 'neutrality' of the performance. He was told that any possibility of the work being considered to be a demonstration of any kind of political view should be avoided. The Weimar city officials believed, if a 'demonstration' was granted permission, all other applications to conduct 'demonstration' (e.g. a neo-Nazi parade) on that monumental square would have to be permitted. Several times he had to clarify that his work was not a demonstration, although he admitted that the work carries many similarities with a demonstration, such as the act of gathering people to stand together on the square (Chong, 2015). He believes that the performance was only granted permission because it was at the end considered to be art not a demonstration – and art is 'innocent' and 'safe' (Chong, 2015). However, he also accredits that such 'devaluation' of art and its political utility does create space for art and its performativity (von Hantelmann, 2014) to both the performers and audiences to emerge (Chong, 2015), through working with the endowed innocent disguise. The 'performativity' that von Hantelmann (2014) is referring to, 'brings into perspective the contingent and elusive realm of impact and effect that art brings about both situationally – that is, in a given spatial and discursive context – and relationally, that is, in relation to a viewer or a public'. Chong (2015) then added to this, by emphasising that:

> the realm of impact and effect in relation to the public addresses changes that refer not only to practical, political and social aspects, but also personal changes that are difficult to evaluate because they are subtle and unclear, and even changing constantly, mostly, referring to construction and understanding of ideologies and body perception.

This complies with my second point of discussion on Chong's performance.

Diverse impacts brought by art: Dissonance between artists' intention and participants' experience/interpretations

While Chong's work was to 'democratise' history by inviting people to tell their personal story or history on the monumental space, which carries a strong normative perception of the authoritarian history and its political denotation, as one of the participants performing in the work, my interpretation of the work went in a rather different direction from Chong's. Remembering the fifteen minutes of performance, I realised that I started with speaking English as I wished to be heard and understood. Then I spoke about my story in the last fifteen years and I tended to justify every decision while mentioning them. At some point, I abandoned speaking English – I started to not want other people to understand me, because I realised that the way I chose which parts of my story to tell does not necessarily bring the whole fifteen minutes of story-telling into a coherent narration. I became selective about what to tell, despite knowing that no one could actually hear me since everyone else on that square was focussing on their own story. My mentality of being selective and seeking a justifiable version of narration overlaps my understanding of how history on all scales is produced, and the intention to be selective so as to acquire a well-rounded narrative is exactly why historiography exists, starting from a national scale and now proven to be relevant on an individual scale.

I cannot deny that such thoughts have a causal link to me being a researcher, but a further disenchantment as such, brought by an experiment of individual scale with my original intention being to show up and support a fellow Hong Kong artist friend – this had to be too random to be predicted and predictable by the artist. This echoes what Chong and von Hantelmann refer to as the 'performativity' (von Hantelmann, 2014) and the impact that performance brings to performers/ audiences in all levels, and through writing them down such impact, in the form of inspiration, hopefully further propagates to readers.

The following section on artistic activism will develop the tension between perceived innocence and the role of performers/participants/audiences in shaping the discursive effects while implementing the art performances. In 2010, I interviewed a group of elderly people in Hong Kong to learn their stories and memories of the Star Ferry and Queen's Piers, which had been demolished. Despite their strong emotions to their site of memories, the group's actions to defend the piers from being removed had been very mild. Most of them in fact did not do anything, and if they did the actions were much restricted to be 'peaceful, rational and non-violent'. The main point from the Weimar example was that bodily movements, be it originally aimed to be art performances or political demonstrations, if it is defined or perceived as art, it becomes innocent, acceptable and penetrating. This is reiterated in the actions taken against Star Ferry and Queen's Pier demolition and the Expressed Rail construction. It shows on the one hand how artists/activists used the innocent disguise of art to attract attention and invite participants by provoking emotions of audiences, and on the other hand, how the implicit artistic actions were more preferable for the politically inactive audiences.

Hong Kong: End of the 'polite' period

From the demolition of Star Ferry Pier in 2006 to the demolition of Queen's Piers in 2007, the transition from the polite modes of protesting in 2006 such as sitting-in, hunger strike and shouting slogans to the direct and bodily actions serves as a paradigm shift. Having in mind the normative and dominant denotation of Star Ferry Pier and Queen's Pier as colonial space, a popular belief was that demolition of the two piers could respond to the quest for decolonisation. This, however, could not stand if one investigates carefully how these colonial sites were related to the anti-colonial history and actions in Hong Kong. In my previous work, I concluded that the removal of the two piers with the forceful measures used by the Hong Kong Government was neo-colonisation instead of any attempt to decolonise Hong Kong (Kam, 2015). The attempt of Hongkongers to defend the piers from being demolished served as 'prelude' of the social unrest in the Pre-Umbrella Movement era. By gloomily losing these two quotidian spaces, Hongkongers stepped on their way of considering their local and colonial history, by questioning what the city, the public space and the non-participatory planning procedures indeed meant to them. The anti-demolition campaigns also served as turning point of how protests were executed before and after these two piers' demolition. Through the frequent and continuous failure of the 'peaceful, rational and non-violent' actions to stop the demolition, new ideas and reflections about modes of protests, law, rules and order as well as Hongkongers' identity emerged, although it was still embryonic at that point.

Ignorance to the local history and hence symbolism of the public space

Lefebvre (1974) delineates the three 'Productions of Space', known as the 'spatial practice' (perceived space), the 'representations of space' (conceived space) and the 'representational space' (lived space). According to Lefebvre, perceived space is affirmed, presumed, produced steadily, controlled and appropriated by society.

Conceived space, produced by architects, planners or designers, is a blend of knowledge, understandings and ideologies. It is conceptual and abstract. Lived space is dynamic, living, interwoven with traces of passion, actions and lived situations which announce its temporal character.

Drawing from Lefebvre's delineations, when positioning the piers as spatial practice or perceived space, the Queen's Pier was perceived as the site for display of colonial power. The pier was infrastructure for British royals and six generations of Hong Kong governors. When they arrived in Hong Kong, they disembarked from their royal yachts at Queen's Pier before putting their 'first step' in the colony. Indeed, the architect and planner, Ronald Phillips, conceptualised the Pier and the Edinburgh Place complex as ceremonial backdrop of the colonisers. With the idea of an axial organisation this colonial infrastructure was created with monumentality. McLehose, one of the most influential colonial governors, put his first step on the colony's pier in 1971. He then walked through the connecting Edinburgh Place, where he reviewed the soldiers in the colonial territory before he proceeded to the City Hall for his inauguration ceremony to assume the governorship of Hong Kong. Cohn (1996) referred to this as colonial power display in anthropological terms. He considered that colonial power was manifested through 'theatrical display, in the form of processions in the pre-modern Europe and elsewhere'. The axial organisation of the piers, the Edinburgh Place and the City Hall delineates the planners' intention to assert and reinforce the monumentality of the cluster of colonial infrastructures alongside with the flow of movement in practical architectural design terms.

If only these first two 'Productions of Space', as defined by Lefebvre, the perceived space and the conceived space, are taken into consideration, conserving the piers in the post-colonial era could be ostensibly problematic. Could one say then, that the demolition of the piers was an act of 'decolonisation'? While considering what Lefebvre categorised as the third 'Production of Space'– the 'lived space' or the 'representational space' – the answer is dissenting. The demolition of the piers was not an act of decolonisation because the Edinburgh complex (comprised of Edinburgh Place, Queen's Pier and Star Ferry Pier), apart from its perceived representation as the colonisers' infrastructure, was in fact the site where the first wave of Hong Kong social movements occurred in the 1960s, according to the newspaper columnist and activist, Chu Hoi Dick (v- artivist, 2007).[1] These movements were primarily against colonial suppressions and were crucial to later societal developments. The inception of the ICAC (Independent Commission Against Corruption) in 1974, which in the last thirty years positioned Hong Kong as one of the least corrupted regions in the world (e, 2016);[2] the official accreditation of the Chinese language from its formerly subordinate position to English (Ervin, Dickson and Cumming, 1998); plus in the society an early discussion on monopoly after the Star Ferry fare topping-up unrest in 1967 (Ngo, 1999). According to Ngo (1999), these societal developments were much connected to the social movements in 1960s; the Edinburgh complex was serving as a backdrop to many of them. Considering how much these listed historical facts marked the milestones in the history of Hong Kong development, they were, however, not included in the history curriculum in school during the colonial era. While 'Chinese History' and 'World History' were obligatory subjects in secondary schools until pupils reached fifteen, local Hong Kong history was not part of the curriculum under colonialism. Without knowing the local history on top of the non-stop eradication of urban heritage – the 'lieux de mémoire' (Nora, 1989) where history and social movements occurred – the lack of knowledge among Hongkongers about their local history implied that there was no information of how political suppressions were fought forty years ago.

During the process of investigating the life and death of the two piers, I spoke to some of the protestors in 2010, two and a half years after their actions defending the piers.[3] Among them was a person named Yu. Yu guarded the Star Ferry's Clock Tower until right before the Clock Tower was demolished. When asked about his intentions in guarding the Clock Tower, the fifty-five year old shared fragments of his stories with me.

Birth and (half-) death 'Wo-Lay-Fay': 'Peaceful, rational and non-violent'

> I was elected, because of being a good pupil, to be the chairperson of the Astronomy Club in school. I was fascinated by the stars in the sky ... every day after school I rode on the tram and arrived here at the City Hall library to borrow books to boost up my astronomy knowledge. When I saw the Clock Tower I knew I should get off the tram. When I heard the chime, I knew that I am here.
>
> (Research interview, 2010, Hong Kong)

Two sets of interviews from two age groups of interviewees were conducted in 2010. One was from a group of elderly people, who were either tour guides or visitors of the Hong Kong History Museum with their age ranging from fifty-two to seventy. They shared with me their memories about the piers, at the same time the reason for their lack of actions during the demolition of the piers. The second group was composed of younger protestors, who did not have much memory attached to the piers, but were the ones who took the comparatively stronger actions when the piers were demolished.

From the elder group, when asked about their stories associated with the piers, they frequently came up with stories showing strong emotions, such as Yu's. However, when asked what they had done to defend their 'site of memory', only Yu out of the six interviewees took part in some actions. However, he only participated up to a point and accused the younger activists of being too vigorous and their actions ineffective. His participation was also restricted. He refused to join any vigorous actions, and his actions to commemorate and guard the Clock Tower included writing emails to the chief executive to petition, taking pictures, writing poems/lyrics and songs to share with the others, talking and showing his newspaper clippings to reporters, helping to organise an exhibition, making 'I-Ching' fortune telling for the Clock Tower's fate, and so on. He considered his intention pure. His actions were aimed at 'dawning on' the government and at the end, even when the piers were demolished then it was also 'okay, although not ideal', as long as he had taken his actions. He was present in most of the incidents and has narrated to me the whole process of how certain other activists started their 'vigorous actions', such as chaining themselves to the entrance of the demolition site and climbing up to the Clock Tower, and he criticised these actions for being ineffective and irrational. His boundary was to stay rational, sensible, peaceful, non-violent and legal.

> In our parent's generations I was brought up with this idea of "just do you own things, learn well and be educated". Like my father he stopped us from joining the campaign for legalising the Chinese language and all other students' demonstrations ... In our generation as university students we had the "4-little-ones concept"– a little car, a little house, a little wife and a little baby. This has been a very pragmatic and materialistic society; everyone turns head to money and benefits, which kills everyone's sense of justice. When a person could not even take care of the money for living, then where does the time for the fighting (for justice) come from? I am not the radical type ... There is just no other way.
>
> (Research interview, Hong Kong, 2010)

A strong sense of 'powerlessness' was displayed to me among most of the older interviewees.[4] However, almost all of them mentioned the Post-80s group[5] and although criticising their actions being 'too vigorous', the elder interviewees were somehow putting their hopes in the young group, hoping that they could perhaps change the stagnant situation of Hong Kong. That was back in 2007.

Direct actions and the metaphorical displays

Chow, in his early thirties when being interviewed, was one of the major activists in conserving the piers from the Post 80s-group I interviewed in 2010. The first time we met he was helping the residents of Choi Yuen Village, who were evicted by the government to relocate due to the controversial Express Railway construction.[6] He was a university lecturer, a farmer and an editor of the 'Independentmedia.net' website.

When being asked to share his personal stories about the piers, his answer was simply negative, accompanied by a big smile. Accused of being 'radical and extreme', his actions included camping on the Queen's Pier for more than a month, organising events and performances, which involved the under-represented citizens taking part in 'metaphorical display'; conducting the rushing-in actions to the already locked-up demolition site of the Clock Tower as well as climbing up onto the bulldozers. He claimed that his objective was purely political; his only intention was to stop the demolition. He took impulsive actions which he considered to be effective at that moment. He considered certain actions such as shooting the last photo or shouting slogans to be despicable and a sole display of powerlessness. He posed a simple yet crucial question: 'If we, Hongkongers – among us with so strong a wish to keep the piers, why don't we do something to directly stop them from being demolished?' He did not draw any boundary to his actions and proclaimed to me that he was psychologically prepared to be arrested by the police and was indeed caught at the end.

In addition to the direct actions aimed at halting the demolitions at the site, Chow and the Post-80 Group also conceptualised artistic performances (which Chow referred as metaphorical displays) with deep reflections on fundamental concepts and colonial history. Drawing from Cohn (1996), if theatrical display was indeed adopted as a grand display of colonial power, Chow and his concept of 'decolonising' the Queen's Pier countered the perceived image of the Pier as 'the colonisers' space' and transformed the Queen's Pier into a site of enunciation of the public. Chow and his companions invited 100 people from various under-represented groups[7] to take a ride on a rented medium-sized fishing boat named 'Localism'. They imitated the disembarking ceremony of the colonisers at the Queen's Pier. The boat-riders arrived at the Queen's Pier and reclaimed the place as not just being 'the Queen's Pier', but also being the Hongkongers' Pier. Flags and decorations were set up to resemble the grand colonial ceremonies from the 1970s.

For Chow and his group, their ultimate objectives were to reflect on the assumptions of actions and the constraints left from the colonial era, as well as the ingrained value of being 'peaceful, rational and non-violent' – which was – and still remains – the major motto and boundary proclaimed by most protestors since the piers' demolition in 2006 and 2007, all the way until 2016, running at times antagonistically to the emerging and growing valiant actions led by the localism belief.

Post-Queen's: Anti-Express Rail Campaign— 'our land, our city'

After the demolition of the two piers, the Post-80s group conceptualised and realised numerous site-specific and ad-hoc art performances and direct actions against various governmental decisions on planning issues. Drawing on and at the same time challenging the ingrained 'peaceful, rational and non-violent' attitude, their actions were seen as cutting-edge and extreme. They pushed the limit of what was still perceived as novel, strange or even radical in 2010. The next action conducted by the group, which consolidated and manifested their objectives, was their protest against the construction of the Express Railway in 2010.[8]

In the interview, Chow admitted to me that all the actions taken in 2010 were aimed at attracting people and touching them emotionally. To arouse sympathy of the public, the idea of 'asceticism' came up. Cheng, another young activist who I interviewed, was one of the organisers as well as the drummer in the ascetic parade/performance. The performance incorporated choreography in some parts. Around 400 young people, dressed in black and white, spent four days walking through five districts – 26 km every day, which equals the distance of the railway segment in the Hong Kong territory. The group conducted the ascetic walk to attract the attention of passers-by and hoped that they would join in along the way, with drumbeats in the background. On the last day, they paraded non-stop around the former Legislative Council's building to show their determination and persistence. During the ascetic walks, they sowed seeds on the ground, and they went down on their knees after every twenty-six slowly proceeded steps. Their performance reminded the audiences, as well as the performers, of their ideal and imaginations to their land while seeds were sowed onto the asphalt road in the Central area of Hong Kong. The group's adoption of artistic performance as actions of protest was aimed at inciting Hongkongers with happiness and ideals rather than a notion of fear or blackmailing, according to Cheng. The concept of 'happy fighting' (Chen, 2010), initiated by Chen in 2010 became the group's driving force when taking social actions. Be it for the participants or for the audiences, they believed that the fight needs to be based on good aspirations rather than fear.

This runs parallel to what Chong (2015) concluded on how 'art' could be perceived as 'innocent', 'safe' and 'harmless' and how such devaluation of art gives potential for art to emerge. Drawing from Chong, it is the devaluation of art which gives potential and a neutralised stage for political participation to emerge, in an implicit way. When a demonstration can be viewed as an art performance, participating in it does not feel as politically explicit. Actions as such succeeded in attracting the originally politically aloof population to, if not to participate, at least to sympathise with the demonstrators who were, with choreography, submissively kneeling on the asphalt road pleading for the authority's mercy.

Forks in the road: 'Peaceful, rational and non-violent' actions to/or valiant actions?

From the Piers movements and the Anti-Express Rail Campaign, the development of different modes of social actions was no longer linear. Tension had been growing in the city since 2007, top-down and bottom-up, introvert and extrovert. Despite the discursive modes of protesting decided by various groups of the few aware Hongkongers to ride on, for the continuously colonised populace, the 'peaceful, rational and non-violent' motto during protesting was still the normative soundbite for many Hongkongers. The rise of localism, later translated by some people into the concept of Hong Kong Independence, guides another path of actions in which activists believe in the direct and corporeal actions.[9] This is, however, not the mainstream mode of taking social actions for the 'peaceful, rational and non-violent' Hongkongers. Social actions are still more marketable if they are laced with artistic and innocent touch. One could seek evidence from the well-known Umbrella Movement. While the international community lauded it as civilised, peaceful and 'blossoming' with creativity, the Umbrella Movement ultimately agitated a semi-aware crowd into action, using artistic actions to cast innocence on strong social quests. At the same time, however, it expelled the ones who remained hard-core believers of 'Rule of Law' being the same as 'obeying the law' – as they believed that occupying public space is in any case illegal so disregarding the purpose of the movement if it is illegal then it should not happen.

Conclusion

For the rather politically aloof Hongkongers in the early post-colonial era, artistic actions were considered acceptable and hence penetrating, especially when actions were considered to be art instead of demonstrations, akin to the Weimarplatz art performance. As the artist Chong accredits, such 'devaluation' of art and its political utility in fact create space for art and its performativity (von Hantelmann, 2014). It allows both the performers and audiences to emerge, with impacts not only practically, politically and socially but also personally, which is difficult to evaluate because of its subtlety, mostly referring to construction and understanding of ideologies as well as body perception (Chong, 2015). For the long-colonised Hong Kong populace, the 'peaceful, rational and non-violent' motto continuously becomes the 'ideal' and the normative boundary for many. Corporeal actions could not and cannot yet become the mainstream of how the populace react to the dissatisfactory authority. The valiant groups are frequently condemned to be violent and their actions 'messing up Hong Kong'.[10] 'Decolonisation is always a violent event' (Fanon, 1963). While the concept of violence considered by Fanon was manifold, the general understanding to the concept of 'violence' limits to the corporeal – violence of the system, verbal, or through political suppressions were seldom addressed in Hong Kong. In the early post-colonial era, when the general populace was accustomed to being obedient citizens, if they had anything to say or to complain about, for example the preservation of the piers, they embedded these 'quests' or 'complaints' in the mildest form such as writing emails to the authority, composing songs, singing them together with other protestors, sitting and shouting slogans and then went home, watching on television with resentment as the bulldozers demolish their everyday space. That was the early form of activism adopted since the invisible climax in 1960s – a segment of local history that was under-represented in school education, hence non-existent in the mind of those who grew up during the colonial era. Later on, during the Queen's Pier conservation in 2007, after the Star Ferry Pier was demolished in 2006, direct actions and explicit artistic performances were put on stage, initiated by activists who were tired of regrets. Protestors rushed into demolition sites, sitting and blocking or even climbing up onto bulldozers. They were heavily criticised by the majority of the others back in 2007. The same group of activists understood the need to bring the 'peaceful, rational and non-violent' Hongkongers into actions by lacing them with emotions and artistic elements. Proceeding to the Anti-Express Railway campaign in 2010, it was a thorough taking off of artistic activism marked by the ascetic parade. It echoed with Chong's idea on 'utilitarianism of art', by making use of the innocent disguise of art to plant impact to the 'situational and relational' audiences. The ideal of 'ultimately fairer society' was advocated and rationalised through the kindness and innocence brought by the perception of art. The Umbrella Movement in 2014 marked an essential period, which emancipated the withdrawn society in terms of enunciation. A 'Hong Kong identity' was created, replayed and reflected: the 'peaceful, rational and non-violent' Umbrella Movement converged the crowds but at the same time urged the reconsideration to the notion *per se*.

The umbrella's symbolism transformed from being quotidian apparatus for weathering conditions to the vulnerable but yet maximum defence one could have under the 87 gas bombs. The Umbrella Movement succeeded in attracting the attention of the politically unengaged crowd with its softened and 'innocent' impression, while political beliefs and ideals were subtly embedded. With the daily fear and threat of being eradicated from the site by the police, the crowd was forced to move on to a reflection to what it means by being 'legal' and 'Rule of Law'. A simple question should be asked by the Hongkongers: how could Hong Kong be considered as a city based on 'Rule of Law' when even the government was breaking the law? The reflection is essential when rationalising the later social unrest such as the 'Fishball Revolution' in 2016. If what protestors were fighting for was justified, then how do we draw the line of the limit of social actions – could discussion of the concept of 'violence' be limited to the corporeal or, should it be extended to the institutional level? Without the accustomed contextualisation and quests to understand nature of matters to its epistemological level, starting with the artistic actions back from the demolition of the piers, I doubt that the aloof Hong Kong populace could achieve today's situation which finally, people debate about government budget, legitimacy and individuals' political beliefs rather than which canton-pop singer to support.

The city can never be decolonised if citizens are inattentive to scope and violation of power possessed by both the authority and themselves. The apolitical post-/neo-colonised Hongkongers needed artistic activism to disintegrate their internalised discipline of being obedient, as well as the sense of being victimised to claim their political beliefs. Hongkongers have arrived at the fork in the road: either capping their quests with the motto – 'peaceful, rational and non-violent' and carrying on being trapped by a cultural imperative of legal conformity, or to be determined, to reflect on the notion of violence and legality to their deepest level, to bring the society forward and fight in the remaining thirty years before a total merge with the world's biggest authoritarian regime, China.

Artistic activism served a great but optimum step to guide the apathetic crowd to the forefront – with the development of concepts such as 'Rule of Law' and 'Localism' brought by artistic activism, the city and its populace ought to depart from the potential brought by it and, to break through the limitations that come with it. 'Decolonisation is always a violent event', as Fanon precisely pointed out in 1963. The emerging localism-oriented politician and social activist Edward Leung argues that for the next protest to win, the fight against political violence must take both paths: the corporeal actions and 'peaceful, rational and non-violent' actions are both essential to make a change.

Notes

1   Chu, Hoi Dick. In a forum in the Edinburgh Place on 29 July 2007.
2   From the Corruption Perceptions Index (CPI), which is published every year by Transparency International (TI) – The Global Coalition against Corruption. In 2014, Hong Kong was ranked seventeenth in the world and Denmark ranked number one. Japan and Singapore are the other two Asian polities ranked higher than Hong Kong.

3 Five narrative interviews were conducted with five candidates with their age more than fifty years old. Two of them were tour guides in the Hong Kong History Museum and three of them were visitors. My aim was to speak to the ones who are interested in the Hong Kong history, and to understand their memory and stories with the two piers.

4 For more of the interviews and analysis please refer to 'Reconfiguration of the Stars and the Queen – A Quest for the Interrelationship between Architecture and Civic Awareness in Post-Colonial Hong Kong' (Kam, 2015).

5 The Post-80 Group, who later on also curated artistic actions in the Anti-Express Rail Campaign, was on the one hand heavily criticised by the elder protestors who thought that their actions were too vigorous, on the other hand regarded as seemingly the hope to change the stagnant Hong Kong back in 2007. Many of them are continuously active in political movements.

6 Until today in 2016 the appropriation of the extra HK$60 billion (about €7 billion) on top of the already used same amount.

7 These under-represented groups were, for example, disabled people, ethnic minorities and victims of urban renewal, also representatives from different political parties, mainly from the leftists.

8 The Railway was proclaimed as aiming to connect Hong Kong to Mainland China but the journey duration, compared to the existing railway system, would not be faster by more than ten minutes after paying the non-capped bill of €14 billion and evicting the farmers from the already rare farmlands. Not to mention that due to legal issues and the fact that the two cities (Shengzhen and Hong Kong) have to maintain their autonomy on law execution according to the Basic Law, the train journey has no possibility to be faster than the current option because of the custom procedures of both sides. Chu Hoi Dick has also been pointing out since 2010, the political intention of constructing such infrastructure was for the Chinese to get ready to transport the communist army and tanks from the Mainland to Hong Kong in case of any motivation of 'Hong Kong independence'.

9 The rise of the localism concept two to three years before the Umbrella Movement is based on the rising tension between the influx of Mainland tourists – whose rights and responsibilities in Hong Kong were much debated, and some Hongkongers, who felt that their everyday living space, culture and dignity were invaded and dispossessed of. Anger and condemns were no longer within the possibility of being hidden in the innocent disguise of any artistic activism. Actions were developed into much direct and expressive. It has seemingly run out of time for the discursive effect of art performance to sublime and develop in the minds of certain performers (protestors) and audiences. The D&G incident in 2012, in which Hongkongers were forbidden photo-shooting at the shop front's display window of the brand's Hong Kong flagship store on Canton Road (the same ban did not apply to the Mainland tourists), aroused tens of thousands of Hongkongers to envelope the shop front, shooting photos while shouting slogans to the shop and requesting an apology. The incident marked the rise of the localism concept, and a growing articulation of it induced the Hong Kong autonomy movement and even a quest of Hong Kong independence, which much in contrast but induced by the 'Wo-Lay-Fay – Peaceful, Rational and Non-violent' Umbrella Movement.

10 Although, it is obvious to many 'peaceful, rational, non-violent' believers that the valiant action takers are at times stating the painful truth and their actions are creating momentary success.

References

Chan, Wan (2010) *Walking Out of the Government Headquarters – Be a Happy Fighter*, Hong Kong: Arcadia Press (originally written in traditional Chinese).

Chong, Wai Isaac (2015) *What Is the Future in the Past and What Is the Past in the Future? Creating Time in Public Space through Performance*, Weimar, Germany: Self-publication.

Cohn, B. S. (1996) *Colonialism and Its Forms of Knowledge: The British in India*, Princeton, NJ: Princeton University Press.

Ervin, G. L., Dickson, P. and Cumming, A. (1998) 'Profiles of language education in 25 countries'. *The Modern Language Journal*, 82 (1), 121.

e, T. I. (2016) *How Corrupt Is Your Country?* Available at www.transparency.org/cpi2014/ results (Accessed: 10 August 2016).

Fanon, Frantz (1963) *The Wretched of the Earth*, New York: Grove Press.

Kam, W. M. L. (2015) *Reconfiguration of the Stars and the Queen – A Quest for the Interrelationship between Architecture and Civic Awareness in Post-Colonial Hong Kong*, Baden-Baden, Germany: Nomos Publication.

Lefebvre, Henri (2009) *The Production of Space*, Malden, MA: Blackwell.

Ngo, Tak-wing (1999) 'Colonialism in Hong Kong revisited'. In T.W. Ngo (Ed.), *Hong Kong's History: State and Society under Colonial Rule*, London: Routledge.

Nora, P. (1989) 'Between memory and history: Les Lieux de Memoire', *Representations*, 26 (1), 7–24.

v- artivist (2007) 729皇后碼頭論壇--新聞所不報的內容系列之十二. Available at www. youtube.com/watch?v=tn-bqQBeZLk (Accessed: 10 August 2016).

Von Hantelmann, Dorothea (2014) 'The experiential turn'. *Living Collections Catalogue*, 1 (1).

9 The climate games

Space, politics and resistance at the COP21 Paris

Ben Parry

Introduction

'One planet, one chance to get it right, and we did it in Paris. We have made history together,' said Christiana Figueres, executive secretary of the United Nations Framework Convention on Climate Change (UNFCCC 2015). The Paris Agreement, instantly dubbed 'the world's greatest diplomatic success' (Harvey 2015) was announced on 12 December 2015, at the UN Climate Conference in a triumphant, self-congratulatory moment of cheering, weeping and embracing between world leaders, UN officials and ministers representing 195 nations. The international press joined in the jubilation, capitalizing on an opportunity for positive news in the political ferment of recent terror attacks in Paris and the ongoing Syrian refugee crisis.

For many critics, scientists and climate activists, this public 'victory' was an overconfident response to non-legally binding commitments to reduce greenhouse gas emissions from fossil fuels and limit global temperature to within a global average rise of 1.5 degrees, without putting any specific policies in place. The significance of the diplomatic success of the Paris Agreement followed failure in Copenhagen in 2009 to draft a new climate treaty intended to replace the Kyoto Protocol. By 2008, when the Kyoto Protocol came into effect, evidence revealed it had failed to slow down global carbon emissions. This was largely attributed to non-ratification of the Protocol by the United States. The Paris Agreement, which comes into effect in 2020, offers a similar deferral of commitment, having generated criticism for having no legal requirement of liability, absence of agricultural proposals and agreeing no fixed financial sums to support developing countries in mitigating against the outcomes of climate change. Sponsored by energy giants, private water companies, automotive, aviation, and agri-corporations including EDF, Suez Environment, Air France, and Renault-Nissan, it is hard to see which of those hard-won gains at the twenty-first Conference of the Parties (COP21) actually sets humanity en route towards zero emissions or a carbon neutral future. Whether or not the agreement is merely symbolic and therefore business as usual in a place of secret trade agreements (Klein 2015), or a genuine step towards leaving fossil fuels in the ground will become apparent in the five yearly reviews, whereby parties agree to publish data on emissions and their progress to limit global warming.

In the wake of the January and November 2015 terrorist attacks in Paris, the COP21 took place under a national state of emergency, and with the city in 'lock down' as part of a nationwide ban on public demonstrations. The simultaneity of these events highlight how COP21 and the terror attacks shaped the subsequent command and control of urban space, media space and the performative spaces of democratic process. These events – of legitimate authority and illegitimate violence – are instances of what Paul Virilio describes as the 'administration of fear' (2012: 15); an environment in which the threat of attacks, natural and manmade disasters (caused by climate change), health epidemics, mass immigration, war and famine combine to make fear a constitutive element of everyday life.

The unfolding of events that surrounded and came to overshadow urgent debates on anthropogenic climate change, set off an alarming performance of politics, power and ideology. Whilst the official delegation of voters rallied for an agreement at COP21, thousands of activists, artists and campaigners on the streets were inscribing unofficial narratives of self-organisation and collectivism in the struggles for climate justice. In revisiting the days surrounding the Climate Summit in Paris, and the wider cultural, political and socio-historical context that shaped events, this chapter examines the role and impact of art within the campaign for climate justice, in particular the convergence of art and activism with the politics of urban space.

This chapter, in part, offers a personal reflection on the events surrounding COP21. At the time of my decision to go to Paris and participate in creative acts of civil disobedience, I had no idea what form my participation as an artist would take. My research interests drew me to explore the role of art within the visual culture of climate change campaigning, with a particular focus on the tools and tactics of art activism and the aesthetic dimension of struggle. To enlist oneself as an artist in the role of struggle is to hold in mind critical questions of art's political efficacy: in what ways does art shape the visual culture of climate change and aesthetics of critique by activists, environmental campaigners, NGOs or grassroots mobilisations? What role can artists play in pressuring governments to commit to limiting the causes of global warming? Can interventionist art help foster an insurrectionary imagination that moves from purely representational to active participation? How can public encounter with acts of civil disobedience, that is physical and embodied, help to activate a critical conscience?

By exploring the circulation of powerful media images and symbols, designed to elicit sympathy and emotion, and highlighting the differing positions and approaches to art and activism within the climate movement, I look at how these actions shape the possibility for critical reflection, and the awakening of a political conscience.

State of emergency

On 13 November 2015, Paris was hit by a deadly wave of terrorist attacks that claimed 129 lives. Within hours the government declared a state of emergency and France closed its borders. In the immediate days that followed, Paris fell silent as people were told to stay indoors. Parks, metro stations, museums, universities and schools were temporarily closed. The streets emptied, except for those moments of collective mourning, alongside the constant presence of international press, police and the military. The impromptu memorials that had appeared outside the Bataclan Theatre and Place de la République grew in size, carrying messages of solidarity, loss, defiance and hope.

The state of emergency had a profound effect on the COP21 and its concomitant climate campaigns in the sense that these events collided in the space of the city. Media attention and public debate was immediately transferred away from climate policy to terrorism and the role of security agencies in protecting the lives of its citizens. The restrictions placed on the movements of predominantly young, white, middle-class activists was set against the enactment of emergency measures in the *banlieues* of Paris, rehearsing long-standing ethnic, race and class divides. Plans for civil disobedience against one form of global threat was now confronted with an emerging spatial politics driven by another: the threat of global terrorist networks and the repercussions of foreign policy in the Middle East (as a driver for which, the West's addiction to fossil fuels need hardly be pointed out).

Under the state of emergency, the French government refused to authorize the peaceful climate marches planned for 29 November and 12 December 2015, to mark the start and end points of COP21. Within activist circles, it was widely known that climate campaigners were to punctuate official meetings, talks and demonstrations with a series of unsanctioned creative and political actions across Paris, and globally. The core of these operations was situated at the state run art space Centrequatre. For five days from 7 to 11 December it hosted a 'Climate Action Zone' (ZAC), which served as a central meeting place for citizen mobilizations and the headquarters of the Climate Coalition 21. Each evening a general assembly was held reporting back from the day's events at COP21, taking place in the Paris suburb of Le Bourget. The ZAC was the main activist information hub, and a space that hosted workshops, screenings, discussions, art exhibitions about climate change and demonstrations of alternative technologies. It was a place 'to increase the rhythm of mobilisations' (Coalition Climate 21, 2015) and connect together the various demonstrations, interventions, political actions and street theatre in the buildup to 12 December 2015, the final day of the Climate Summit.

When the state of emergency was extended in advance of the climate summit by an additional three months, controversy over the right to peaceful demonstration grew in the media. Many human rights groups and activists warned of the threat to fundamental freedoms and civil liberties (Chrisafis 2015). In anticipation of anti-COP21 activities, French police carried out raids on squats, homes and locations of known activists, seizing computers and personal effects (Nelson 2015), and placed twenty-four climate activists under house arrest, accusing them of flaunting the ban on protest under the state of emergency. The targeting of activists, widely reported by the mass media (Clarke 2015; Chrisafis 2015; Nelson 2015) provided a platform for campaigners who were able to challenge detention and criticize the curtailment of civil liberties. In sharp contrast, hundreds of Muslims and immigrants in the suburbs of Paris and across France, who as victims of anti-terrorist laws experienced violent raids, arrests and detentions were not afforded the same profile in the mass media. Spatial divides and socio-economic

Figure 9.1 Naomi Klein speaking to 3,000 activists at the Climate Action Zone, Centrequatre, Paris [photograph: Ben Parry].

marginalisation further exacerbated their exclusion. In other words, as 'differentially empowered publics' (Warner 2002) they encountered power struggles at the bottom of a hierarchy of access to the public sphere and modes of publicity. Their ability to engage in peaceful political protest was further restricted by oppressive policing of France's border regime 'as the absolute command over the distribution of space' (Raunig 2007: 251). On the ground, the nature of these security measures served to sharpen my focus on the *right to the city* and how that might be interpreted differently for each of us. These momentary yet perennial exclusions thus highlighted the universal mandate of public space: to be open to all. As Lefebvre explains, the right to the city

> signifies the right of citizens and city dwellers, and of groups they (on the basis of social relations) constitute, to appear on all the networks and circuits of communication, information and exchange. This depends neither upon urbanistic ideology, nor upon an architectural intervention, but upon an essential quality or property of urban space: centrality. Here and elsewhere we assert that there is no urban reality without a centre, without a gathering together of all that is born in space and can be produced in it, without an encounter, actual or possible, of all "objects" and "subjects".
>
> (1996: 195)

Due to the long period of planning and intense organization behind direct action and street theatre, artists and activists tended to overlook our privileged position of centrality in a rehearsed peripheralisation. The added complexity under a state of emergency, demanding more focus, discussion and self-determination, only served to distance our neglect of the social and spatial exclusion of the *other*, now driven by anti-immigrant sentiment. The attempted reclamation of the space of the city by activists, restricted under state of emergency, neither afforded nor bargained for the struggles of France's immigrant populations. This highlights the lack of a pluralist perspective of group demographics within the movement for climate justice, and articulations of radical democracy more broadly, where the representation of marginalized groups, in particular populations and countries most at risk must surely be empowered to speak, challenge and be present in all networks. For members of immigrant communities living in the *banlieues*, the social relations and conditions necessary for the formation of political subjectivity (the extent of an individual's participation in social and political change) are dependent on their ability to challenge social, cultural, economic and racial practices of exclusion. To unpack this a little further I must turn to a short history of the state of emergency and its relations to the struggles of the colonized.

Decolonisation and modernisation

Before 2015, the state of emergency was last issued in France during the riots of 2005 that stemmed from violent confrontations between police and immigrant youth placed under curfew in the Paris suburbs of Clichy Sous-Bois. Fuelled by aggressive police tactics, the riots quickly spread to cities across France that lasted two weeks (Hussey 2014). This, the largest eruption of civil unrest in France since 1968, was the outcome of prolonged alienation and social exclusion in the suburban *banlieues* that remains rooted in the legacy of colonialism.

The combined threat of random acts of terrorism with police repression of migrant communities in the *banlieues* and on the streets of Paris in 2015, share troubling elements with the environment of fear created during the final years of the Algerian War of Independence (1954–1962). In the early 1960s, the city became the scene of reprisal attacks, bombings and assassinations between supporters of the National Liberation Front (FLN) and far-right groups such as the Secret Armed Organisation (OAS) seeking to prevent Algeria's independence from French colonial rule. On 26 April 1961, after a failed *coup d'etat* in Algiers threatened to erupt in Paris, Charles de Gaulle decreed a National State of Emergency across France that persisted, on and off, until May 1963. The state of emergency, introduced for the first time, had allowed for increased state violence, counterinsurgency measures, a total ban on all public demonstrations and protest, the systematic use of torture by French forces and widespread police brutality with curfews and raids carried out on Algerians living in what were then immigrant worker shanty towns and workers' hostels in the outskirts of Paris.

'The immigration that haunts the collective fantasies of the French today,' according to Kirsten Ross 'is the old accomplice to the accelerated growth of French society in 1950's and 60's' (2002). In her book, *1968 and Its Afterlives,* Ross described the separation and suppression of the narrative of decolonization from the period of French modernization after World War II. This period of decolonization, of mass migration from North Africa into Europe, was also the great post-war movement in urbanization; when 'urbanism becomes ideology' and the 'moment the historic city exploded into peripherics' (Lefevbre 1983). In a massive spatial reworking of the social boundaries of French cities (Pinder 2005), immigrants and in particular French Algerians, employed in construction and manufacturing industries, were ghettoized in the new *banlieues* of Paris (Sadler 1999: 52). This ran in tandem with the architectural theorists of the day espousing the utopian ideals of Le Corbusier's brutalist style of mass housing developments.

By reframing the simultaneity of decolonization and modernization with the socio-political events of 1968, Ross suggests that the coming together of students with workers during the May uprising was unthinkable if not for the involvement of civilians in protests during the end of the Algerian war. On 17 October 1961, 30,000 pro-independence Algerians staged a peaceful protest on the streets of Paris against a racist curfew. Attempting to enforce the ban, police killed an estimated 200 protesters. In the aftermath of the police massacre of Algerians, a political awakening spread through the French public, that encompassed a durable radicalisation within the student milieu that led to the first street demonstration against the war (*Situationist International*, 1968). This marked 'the first instance of an intervention at the national level in a new mode, of students as a political force, for a cause that was not a defence of student interests' (Ross 2002: 56). Ross argues that it was this transformation of political subjectivity, established through allegiance and solidarity with the *other* whilst refusing identification with the actions of the state, that enabled students in 1968 to identify with workers and engage in struggles outside their own social identification (Ross 2002: 83). As a student at the time, Jacques Rancière also recalls the formation of new forms of political subjectivity during the French-Algerian war.

> This political subjectivation was primarily the result of a dis-identification with the French state that had done this in our name and removed it from our view. We could not identify with the Algerians who appeared as demonstrators within the French public space, and who then disappeared. We could, on the other hand, reject our identification with the state that had killed them and removed them from all the Statistics.
>
> (Rancière 1998)

On that fateful day, police threw beaten and unconscious bodies into the Seine that were washing up on its banks for several days. In response, an insertion of political graffiti suddenly appeared on the Saint Michelle Bridge that read: *'Ici on noie les Algériens'* – 'Here we drown Algerians'. Fifty years later in 2011, an anonymous activist re-sprayed 'Here we drown Algerians' at the original site. Had this been done during the climate summit, or during the prolongation of the state of emergency, it would have not only spoken of France's long struggle with integration, but of the reenactment of raids and curfews being carried out against immigrant populations.

An earlier instance of the 'dissidentification' described by Rancière appeared in September 1960, in the form of a published manifesto, 'Declaration on the

Right to Insubordination in the War in Algeria'. Also known as 'the Manifesto of the 121', the censored text had been written in response to the pursuit, imprisonment and sentencing of those refusing to participate in the war, and for actively supporting Algerian Independence. 'Outside of pre-established frameworks and orders, by a spontaneous act of conscience, once again a resistance is born; seeking and inventing forms of action and means of struggle in a new situation ...' (Blanchot 1960). The 121 artists, intellectuals, poets and filmmakers that signed the manifesto (including Jean Paul Sartre, Simone de Beauvoir, Francois Sagan, Andre Breton, Alain Resnais, Guy Debord and Florence Malraux) were met with heavy penalties and sanctions in an 'open war against all cultural freedoms' (SI 1960), including a suspension of educational jobs and ban on all public presentations of film, theatre, radio and television.

The second major incident occurred on 8 February 1962, during a peaceful protest organized by students against the war and the escalating criminal acts of the OAS,[1] which resulted in tragedy when police forced large crowds into Charonne Metro station and crushed eight people to death. Five days later an impromptu crowd of over 500,000 people gathered to honour the victims of Charonne and attend the funeral at Père Lachaise Cemetery. This became the first of a series of leftist funeral rituals[2] in the streets of Paris, witnessed again in 1986 when a procession of 600,000 people attended the funeral of French-Algerian student Malik Oussekine, who was killed in police custody following his arrest during student demonstrations against university reforms and immigration restrictions.

Such ritual, not seen in recent times until the *Charlie Hebdo* attacks of 2015 in which seventeen people were killed, prompted 1.5 million Parisians, joined by world leaders, to form a procession through the streets.[3] Exclusion from the system was visible by the absence of large numbers of immigrants from these crowds. Hélène Kuhnmunch, a history teacher in the *banlieue* north west of Paris, remarked that she did not attend the solidarity march, as she knew 'that the *banlieues* would not be there' (cited in Packer 2015). Under the state of emergency, the ritual of public mourning became legitimized public expression, and peaceful protest took the form of a unitary co-option of social solidarity by the state. Put simply by Augusto Boal, 'Now, Freedom must be supervised' (Boal 2000: xi). In the immediate aftermath of the *Charlie Hebdo* attacks, the state spoke out in defence of freedom of expression as one of the cornerstones of French values, exported globally via the mimetic slogan *Je Suis Charlie*. However, these legitimated moments of public expression help mask the infringements to civil liberties under a state of emergency. 'This leads us to a violent paradox,' warns Paul Alliès (2015): 'the institutionalization of exceptional powers reduces the scope of public liberties even as it is justified by the need to defend those liberties.'

Circulation of media symbols and a 'democracy of emotions'

Just hours after the November 2015 terror attacks, graphic designer Jean Julian created the symbol 'Peace for Paris' in a sleight-of-hand modification of the original Campaign for Nuclear Disarmament (CND) sign to incorporate the Eiffel Tower.

This symbol also went viral and appeared in various forms across the world, on posters held up at campaigns, vigils and expressions of solidarity, stencilled on t-shirts, on screens at sports events, projected onto buildings and across social media feeds. This rapid response perfectly captures the utopian ideals of John Fiske's 'semiotic democracy', in which people are able to 'recode' the projected meanings and signifiers of popular culture, commandeer systems of circulation and 'rework' cultural commodities into new languages of critique and struggle (Fiske 1989). In a semiotic democracy, the once passive consumer is now able to disrupt the ideological circuits of the producer (Meireles 1994), empowering individuals 'to add to the rich and expansive cultural fabric of a true public domain, where everyone participates in the ongoing process of cultural production' (Katyal 2012: 50).

Certainly, such moments demonstrate how social media has enabled broad participation. Yet the apparent removal of certain barriers, divisions and distances formed in these phantom emulations of togetherness and affinity, masks a world in which the gulf of inequality and access to resources has never been so pronounced and exclusionary. In these mediated expressions of collective identity and struggle – against an invisible enemy – one also senses a convenient displacement of antagonisms in favour of mass consensus. In the same way, the political antagonisms portrayed in the satirical pages of *Charlie Hebdo*, liked then only by its 30,000-strong left-liberal readership (which subsequently briefly rose to 300,000) were reduced to a collective symbol of free speech, holding aloft a pen. As isolated media events, given the constant incidence of terrorist atrocities from Turkey to Mali that barely make the back pages of Western newspapers, the public response to Paris's recent attacks are better understood as isolated media events, that signify a shift in approach to 'manufacturing consent' from that of opinion to emotion.

By reviewing the events surrounding the COP21 through the lens of decolonization and securitization under the state of emergency, it becomes possible to observe a shift in political subjectivity within the campaign for climate justice itself. This oscillates between two political poles; on one side, a position of counter-power and a refusal to negotiate with the state, and on the other, a re-identification with state power and security ideology. The unfortunate combination of the COP21 and the terror attacks, as globally mediated events, highlight a shift in the mechanisms of democratic process and modes of 'networked consensus' – a process now enabled by the real-time transmission of events – which makes possible 'the synchronization of emotion on a global scale' (Virilio 2012: 13). The terrorist attacks in Paris, for example, generated a situation in which 'the same feeling of terror can be felt in all corners of the world at the same time. It is not a localized bomb: it explodes each second, with the news of an attack, a natural disaster … it creates a community of emotions' (ibid). The administration of fear then acts as a form of command and control of urban space, creating consensus in media space through a synchronization of emotions that must first express empathy and solidarity over any language of opposition or critique.

This subtle shift in political process is understood in Virilio's terms as the transition from a 'democracy of opinion', to a 'democracy of emotion' (2012: 131). This can be seen clearly in the mass appropriation of the response 'Je Suis Charlie' and 'Paris for Peace', as millions superimposed the colours of the Tricoleur on their social media profiles in an expression of solidarity. The shift in the nature of this public response can be attributed to both the instantaneity of real-time communications and our rapid switching through digital information as a state of permanent distractedness that tends to collapse reflection in favour of 'automatic responses'.

> The digitized world has no respect for contemplation or reflection; it delivers instant stimulation and gratification, forcing the brain to give most attention to short-term decisions and reactions ... There is a move away from a society made up individuals with distinct combinations of knowledge, experience and learning to one in which most people have socially constructed, rapidly acquired views that are superficial and veer towards group approval rather than originality and creativity.
>
> (Standing 2014: 32)

My concern here is that the visual culture of climate change also reveals a deeper reliance on the power of emotional persuasion over reasoned argument, through attempts to condense our impact on the natural world into a single powerful image. (I am of thinking of the polar bear, better still the polar bear stranded atop melting ice.) Such images, be it the poster species of global warming or the techno-scientific images taken from space, Such attempts at encapsulation, argues Emily Scott, be it the poster species of global warming or the techno-scientific images taken from space, are simply too literal to contend with the wider systemic crises of climate change; 'both the on-the-ground embedded view and the global view from afar tend to curtail political agency' (Scott 2016: 132).

On the whole, questions of activist art's efficacy are still beholden to the magnitude of media dissemination of the image-event, with an over-reliance, therefore, on the power of osmosis. If an image 'goes viral', so the argument goes, its 'efficacy' is seen to be worthwhile. Under these conditions, if these interventions and acts of civil disobedience are image-events consumed as spectacle, can they generate more than short-term reactions?

On the eve of the Climate Summit, up to 400,000 Parisians were expected to take part in the global movement for climate justice, which saw over 700,000 people in 175 countries join in peaceful demonstrations. They were to congregate at Place de la République, now the symbolic site of memorial and spontaneous vigil following the *Charlie Hebdo* attacks in January that year. At its centre, a statue of Marianne, figure of liberty and symbol of the French Republic holding aloft a tablet of the Declaration of the Rights of Man and Citizen, and surrounded with three statues personifying liberty, equality and fraternity. On the morning of 29 November, in response to the ban on demonstrations throughout France, a spontaneous assembly of 10,000 pairs of shoes appeared in line formation across Place de La Republique in place of the missing protesters denied freedom of speech and freedom of assembly (see Figure 9.2). This was one of the few actions of the climate campaign that successfully captured the collision of both events. The poetic gesture that respected the silence of the square, created presence of absence through a powerful visual image of creative resistance and solidarity. These creative acts become important

Figure 9.2 10,000 shoes, Place de La Republique, Paris, 29 November, 2015. [photograph: John Englart].

symbols and languages of critique, much like 'the people's piano' that appeared in front of police lines in Kiev, which quickly went viral.

By late afternoon, 400 protestors gathered near the statue were met by climate activists, who in defiance of the ban had formed a human-chain stretching down the adjoining Boulevard Voltaire. As police moved in to clear the square, the situation quickly escalated into clashes with protestors, who promptly deployed elements from the impromptu memorial as projectiles. Other members of the crowd moved to circle and protect the shrine, many of whom were pushed and dragged by police through the flowers, glass candles, written notes and flags. The image of 10,000 ranked shoes was thus replaced in the media with one of police riot shields, pepper spray and tear gas and the remnants of public mourning scattered and broken across the square. After 200 arrests, the state quickly restored its authority and the tone for the rest of the Climate Summit was set.

The scale of the police and military presence created a sense of unease for activists. Tensions were high. Any acts of civil disobedience, trespassing and appropriation of physical space and media space now carried the threat of severe penalties. Climate activists were waiting to see what happened next. Word spread of the sudden appearance of a perfectly executed city-wide action by the clandestine UK based organization Brandalism. Brandalists combine neo-Situationist urban tactics with media *detournment*, subverting of marketing slogans and signs in a front-line war against political obfuscation and commodity distraction. On 25 November small teams across the city took over 600 advertising sites. In coordinated groups, they replaced the posters at bus stop shelters with *detourned*

Figure 9.3 Brandalism, bus stop ad takeover. *Etat d'urgence*. Artwork by Eubé, Paris, 29 November 2015 [photograph: courtesy of Brandalism].

images by over eighty artists (see Figures 9.3 and 9.4). Their targets for corporate takeover were the greenwashing advertisements by various energy giants, banks, airlines and car manufacturers sponsoring COP21. 'Greenwashing' is a process in which deceptive environmental marketing campaigns promote misleading practices, policies and products, perpetrated by companies attempting to clean up or 'whitewash' their public image. Despite a police raid on Le Annex squat where the posters were being made, just hours before crews were due to head out, the project had slipped the authorities' net.

When the Brandalism project finally broke news, the project attracted headlines in broadsheets, with the bus-stop posters becoming the illustrative image of greenwashing. The flood of global media was unexpectedly celebratory with journalists across the web expressing admiration of how, under a state of emergency, this scale of appropriation of advertising space could be pulled off. Brandalism perfectly illustrates Sonia Katyal's purposing of the term 'semiotic disobedience' to describe artistic practices that 'involve the conscious and deliberate re-creation and occupation of property through appropriative and expressive acts that consciously risk violating the law that governs intellectual or tangible property' (2012: 52). Unlike semiotic democracy, argues Katyal,

> which seeks access to symbols, semiotic disobedience seems to take issue with the very idea of ownership of symbols entirely. In this way, the tactics utilized by semiotic disobedience activists offer an interesting convergence of property and speech by targeting – and challenging – the "sovereignty" of advertising.

(2012: 60)

Figure 9.4 Brandalism, bus stop ad takeover, *Solution 21* by Bill Posters, Paris, December 2015 [photograph: courtesy of Brandalism].

For the climate movement, it had the much-needed effect of restoring civic courage and reigniting the fires of civil disobedience. The action also illustrated the point that validation by the media is often of key importance to the perception of 'success' for performers, activists and cultural hijackers. As Richard Schechner suggests, 'the Street is the Stage', and they need this legitimation 'in order to feel that the event is real … It is most theatrical at the cusp where the street show meets the media' (1993: 51). When Steven Durland (1989) introduced the visual and theatrical actions of Greenpeace (est. Vancouver 1971) into the fold of critical public art practices, he outlined the essential relationship between protest actions and use of images in the form of 'guerrilla theatre'. Greenpeace was instrumental in relaying the importance of creating a striking visual image or 'mind bomb' that could switch consciousness (Hunter 2015). Abbie Hoffman wrote: 'Highly visual images would become news, and rumor-mongers would rush to spread the excited word' (Raven 1989). Creating 'image-events' in the media was crucial to the power of distribution that remains vital to the practice of interventionists working today, and Greenpeace made one of the most spectacular and unexpected contributions to the climate actions in Paris. Whilst circling the Arc de Triomphe, several trucks lifted their tailgates, whilst a team of cyclists pulled the plugs from jerry can panniers, dumping thousands of litres of yellow paint onto the road. Within seconds, cars had spread the paint not only around but down the adjoining roads, fanning out to create a giant yellow sun (see Figure 9.5). Whilst others scaled the monument to hang the customary Greenpeace banners, the police rounded up over 60 activists on the ground. This action in particular highlights the differing positions and attitudes towards the relationship between art and activism. It raises questions about, on the one hand, the

Figure 9.5 Greenpeace, giant sun surrounds Arc de Triomphe as climate talks approach
end, Paris, 11 December 2015 [photograph: ©Greenpeace].

artistic need for ambiguity, and on the other, the more didactic and symbolic nature
of direct action and protest, and how these may or may not open up possibilities for
critical reflection, and the awakening of political conscience.

Climate Games: Where action-adventure meets actual change

Despite the ban on protest and the threat of police violence, the vast plans of the
climate coalition went ahead. Many of these creative actions were drawn together
under the banner of the 'The Climate Games', the main event of which would be
the *Red Lines* action of mass civil disobedience on 12 December 2015. The Climate
Action Zone hosted at Centrequatre was just one of many centres of self-organisation in a citywide network of spaces in a well-planned infrastructure, set up to support and house activists, facilitate actions and build capacity within the movement
(see Figure 9.1). These included squats, solidarity centres, trade union centres and
art spaces that were turned over to hostels, meeting houses, workshops and production facilities.[4] One of the most significant aspects of self-organisation within the
movement and a key part of its infrastructure came courtesy of the *Anticop United
Kitchen Front*; made up of people's kitchens from all over Europe that provided
tens of thousands of free meals to organisers and participants each day, stationed at
all known locations, and popping up at mobilisations, on demos and after-parties.

Art, activism and the right to the city converged in The Climate Games as activists, artists, coders, designers, writers and performers created interventions, inter-
ruptions and acts of civil disobedience across Paris, France, and globally. Over

fourteen days, from 29 November to 12 December 'whilst the UN summit played with words', 214 actions by 124 teams 'demonstrated how disobedience is the mother of creativity within the Climate Justice movements' (Climate Games 2015). Each participant uploaded their project to the website, geo-tagged their location on a map that showed the distribution across the city. Activists targeted conference sponsors and fossil fuel polluters, storming the bank lobbies of BNP Paribas; indigenous kayakers from the Americas, Indonesia and the Congo paddled down the Seine demanding secure land rights and protection of waters; a Scandinavian cycling outfit visited the Volkswagen showroom to present their revolutionary new product; molasses was spilled across the floor of the Louvre calling for the museum to drop its oil company sponsors, and the leading scientific reports on climate change prepared for governments by the Intergovernmental Panel on Climate Change (IPCC) were printed onto toilet rolls that furnished the conference bathrooms at Le Bourget. These are just some of the many inventive approaches that gave a sense of carnival occupation of the climate debates during COP21.

The premise of The Climate Games was to connect disparate actions, so that instead of existing in isolation of one another they instead formed a concatenation of acts of resistance. This created an extended sense of carnival, active on all networks, creating embodied and spatial encounters with dissent, disobedience, urban play and the counter hegemonic narratives of social movements. Linking together tactical approaches to physical actions and embodied responses, such as those taking place under the banner of The Climate Games, combines the strategic thinking and infrastructure of the wider movement. The civil disobedience of The Climate Games operated within the oppositional space of the strategy and tactic, a hybrid space that resisted the imposed spatial order whilst possessing an immanent capacity to engage those organizing systems in alternative ways of operating. As such, these actions aspire to build a more effective resistance than the tradition of the mind bomb, that in an image-saturated world may offer only a fleeting epiphany and an over reliance on the power of osmosis. The model of The Climate Games is therefore to provoke a rethinking of the possibilities of a joined-up approach to art-action and of the importance of the body through physical participation. A hybrid of tactic and strategy might be termed 'tictactic' or the pursuit of 'tacketry', (Parry 2012) a term imagined to imply the (potentially endless) concatenation of a chain of tactics – an emergent politics linked to this, like the co-ordination of disparate tactical activity as seen in the Arab uprisings, which gives form to the concept of a 'concatenation of resistance' practices (Raunig 2007).

Do not cross the Red Lines

The red lines are the minimal necessities for a just and liveable planet. Red lines that must never be crossed.

On 12 December, with Paris still in a state of emergency, 15,000 people formed *Red Lines* (people dressed in red, holding red tulips, carrying giant red banners and red umbrellas) along Avenue de La Grande Armée, between the Arc de Triomphe and Porte Maillot (see Figure 9.6). At the same moment as the red lines moved through the streets to gather in front of the Eiffel Tower and Champs de

Figure 9.6 The Red Lines Protest, Avenue de La Grande Armée, Paris, 12 December.
[photograph: John Englart].

Mar, delegates were adopting the Paris Agreement to combat climate change. The red lines were a statement of co-presence, so that rather than the delegates at Le Bourget claiming the future agreement of climate change, it is 'We who will have the last word' on climate change and 'that word will be written with our disobedient bodies' (Climate Games 2015).

Prior to the state of emergency, the intention of *The Red Lines* was to have formed a giant human chain, circling Le Bourget, but the terror attacks had transformed the summit into the most heavily securitised space in and around the capital, making it impossible for non-authorised persons to get within 100 yards of the venue. Modified plans about where the red lines action would take place were supposed to be kept secret until just hours before mobilisation. But in the final days of preparation, everything threatened to collapse under intense negotiations with police and amongst the coalition of climate campaigners, divided between the prospect of state endorsement and a refusal to negotiate the action of mass civil disobedience. In the end, we were forced to disclose our intentions to form red lines extending out from the Arc de Triomphe, that were to be acted out regardless of state interference. Attempts to stop *The Red Lines* would produce media images of violent clashes with police, suggestive of a loss of state control, or highlighting a break with an affective democracy of emotions. The police were therefore disposed to allow the event to proceed. They maintained comparatively low numbers, relaxing their blockades when the red lines of disobedient bodies moved through the streets to the foot of the Eiffel Tower. Whilst joining the red lines at the specified location, I also observed how the majority of police, in larger numbers, were stationed around the approaches to the Arc de Triomphe. This was a subtle move less to keep us under control, but more to prevent impromptu participation, intimidating members of the general public who might be tempted to join our lines.

Conclusions and openings

By viewing the anti-COP21 actions and the wider public response to the state of emergency through the lens of decolonization and securitization, we can observe a complex process of formation and deformation of political subjectivity, generated both by the response to terror attacks and the campaign for climate justice. Had the attacks not taken place, the public might have been more attuned to positions of counter-power, and adopted a more critical position towards government cooperation with energy corporations and trade agreements, through a process of disidentification. Instead, the administration of fear catalysed a necessary re-identification with state power and security ideology.

An over-reliance on emotional imagery and didactic slogans in climate campaigning avoids addressing the question of how political subjectivity and a critical conscience is formed. Within the discourse on art and activism, too few questions are being asked of the processes of political transformation from passive consumers to becoming active in struggle, the relations between 'those who look and those who act' (Rancière 2010: 19). What exactly catalyses a political awakening, making the transition from spectator to actor in political struggle? I have outlined the importance of creative disruption that can function equally as well through a position of ambiguity, rather than the didactic aestheticization of politics as favoured by working groups within social movements, which aim for maximum media coverage. The shortcomings of this narrow approach to art's potential efficacy, falls prey to the trap of how we consume spectacle in the twenty-first century, and our momentary attention span or passive reception of said images.

When Brecht created his 'Epic' theatre, he wanted his audience to cease being passive spectators and undergo political transformation. Through the use of humour, disruption and insertion, Brecht highlighted the constructed nature of event as theatre, so that the audience could move beyond an emotional identification with the subject and adopt a critical position instead. In this way, the spectator would exit the theatre and take action to affect change in the real world. Agosto Boal believed that 'to stimulate the spectator to transform his society, to engage in revolutionary action' one needed to seek another kind of poetics. And that this should be based on the principle of trespass, an invasion of the scene in which the spectator appropriates the power of the actor. So, that for Boal, 'the audience mustn't just liberate its Critical Conscience, but its body too' (2000). Acts of civil disobedience during the COP21 and beyond, offer some examples of moments when insurrectionary imagination moves from purely representational to active participation, when we become actors in struggle against climate chaos, when our bodies become essential to the 'dynamics of resistance' (Klein 2015a: 450)

Without an embodied experience of the symbolic trespass, a critical conscience like a fleeting epiphany cannot develop a durable political awakening to actually transform our patterns of behaviour. The urgent debate now concerns itself with how to combat and overturn a democracy of emotions and short-term reactions, towards a political transformation as an embodied and enduring participation in both one's immediate living environment and broader social struggles.

The fertile collaborations between artists and activists continue to bring new poetics and objectives into social struggles, to invent new forms of expression of theoretical thought and practical action against the destruction of a civilization based on the existence of the commons. As Pierre Bourdieu makes clear: 'Our objective is not only to invent responses, but to invent a way of inventing responses, to invent a new form of organization of the work of contestation, of the task of activism (Bourdieu 1988: 58).

Notes

1 The demonstrations were sparked by a series of attempted target killings of political, trade union, academic and literary figures, including the failed bombing of André Malraux that blinded a four-year-old girl in his apartment building.
2 A ritual that continued up until the death of Jean-Paul Sartre in 1980 in which 50,000 joined his funeral procession through the streets to Montparnasse cemetery (Ross: 83).
3 Including British Prime Minister David Cameron, German Chancellor Angela Merkel, Spanish Prime Minister Mariano Rajoy, Israeli Prime Minister Benjamin Netanyahu and the Palestinian Authority leader Mahmoud Abbas.
4 Core participating spaces include 'Bourse du travail' on Place de la République; the 'Bourse du travail' of Saint Denis; The Jardin D'Alice in Montreuil; Le Annex; the Centre international de culture populaire (CICP) in the 11th arrondissement; Salle Olympe de Gouge (11th arrondissement) and the Salle Jean Dame (2nd arrondissement).

References

Alliès, P. (2015) *A Short History of the State of Emergency*. Verso [Online]. Available at www.versobooks.com/blogs/2349-a-short-history-of-the-state-of-emergency-by-paul-allies [Accessed: 22 November 2015].

Blanchot, M. *et al.* (1960) 'Declaration on the Right to Insubordination in the War in Algeria' ('The Manifesto of the 121'). Available at www.marxists.org/history/france/algerian-war/1960/manifesto-121.htm CopyLeft, Creative Commons (Attribute & ShareAlike) marxists.org 2004.

Boal, A. (2000) *Theatre of the Oppressed* (Third Edition), London: Pluto Press.

Bourdieu, P. (1988) *Acts of Resistance, Against the New Myths of Our Time*, Cambridge, UK: Polity Press.

Calvino, I. (1974) *Invisible Cities*, London: Harcourt.

Chrisafis, A. (2015) 'France's state of emergency could lead to abuses, say human rights groups'. *The Guardian* [online]. Available at www.theguardian.com/world/2015/nov/26/frances-state-of-emergency-could-lead-to-abuses-human-rights-groups-warn [Accessed: 14 December 2015].

Clarke, M. (2015) 'COP21: Security crackdown in Paris sees climate change protesters under house arrest' [online]. Available at www.abc.net.au/news/2015-11-29/climate-protesters-banned-in-paris-security-crackdown/6983870 [Accessed: 2 December 2015].

Climate Games (2015) 'Climate Games, we are nature defending itself' [online]. Available at www.climategames.net/home [Accessed: 5 November 2015].

Coalition Climate 21. (2015) 'Climate Action Zone' [online]. Available at coalitionclimat21.org/en/climate-action-zone [Accessed: 30 November 2015].

Fiske, J., 1989. 'Moments of Television: Neither the Text Nor the Audience'. In Seiter, E., Borchers, H., Kreutzner, G. and Warth, E.-M. (eds), *Remote Control: Television, Audiences, and Cultural Power*, London: Routledge, pp. 56–78.

Gore, A. (2015) 'Speaking with one voice to solve the climate crisis'. *The Huffington Post* [online]. Available www.huffingtonpost.com/al-gore/speaking-with-one-voice-to-solve-the-climate-crisis_b_8794402.html [Accessed: 14 December. 2015].

Harvey, F. (2015) 'Paris climate change agreement: the world's greatest diplomatic success'. *The Guardian* [online]. Available at www.theguardian.com/environment/2015/dec/13/paris-climate-deal-cop-diplomacy-developing-united-nations [Accessed: 14 December 2015].

Hussey, A. (2014) *The French Intifada: The Long War Between France and Its Arabs*, New York: Farrar, Straus and Giroux.

Hunter, E. (2015) 'Then & Now: Launching a "Mind Bomb" to save the Arctic' [online]. Available at www.greenpeace.org/international/en/news/Blogs/makingwaves/save-the-arctic-mind-bomb/blog/53301/ [Accessed: 11 January 2016].

Katyal, S. K. (2012) 'Between semiotic democracy and disobedience: Two views of branding, culture and intellectual property'. *4 Wipo J. Intell. Prop.* 50, 60. Available at ir.lawnet.fordham.edu/faculty_scholarship/618.

Klein, N. (2015) 'COP21 Climate Emergency. Speaker at the Climate Action Zone'. Paris, France: Centrequatre, 104.

Klein, N. (2015a) *This Changes Everything*, London: Penguin.

Lefebvre, H. (1983) 'Henri Lefebvre on the Situationist International'. Interview conducted and translated in 1983 by Kristin Ross [online]. Available www.notbored.org/lefebvre-interview.html [Accessed: 2 Feburary 2016].

Lefebvre, H. (1996) *Writings on Cities*, (trans.) Kofman, E. and Lebas, E. Oxford, UK: Blackwell Publishing.

Meireles, C. (1994) 'Notes on Insertions into Ideological Circuits 1970–75'. In Doherty, C. *Situations*. London, Cambridge: Whitechapel Gallery & MIT Press.

Nelson, A. (2015) 'Paris climate activists put under house arrest using emergency laws', *The Guardian* [online]. Available at www.theguardian.com/environment/2015/nov/27/paris-climate-activists-put-under-house-arrest-using-emergency-laws [Accessed: 28 November 2015].

Packer, G. (2015) 'The Other France: Are the suburbs of Paris incubators for terrorism?' *The New Yorker* [online]. Available at www.newyorker.com/magazine/2015/08/31/the-other-france [Accessed: 7 January 2016].

Parry, B. (ed.) (2012) *Cultural Hijack: Rethinking Intervention*, Liverpool, UK: Liverpool University Press.

Pinder, D. (2005) *Visions of the City. Utopianism, Power and Politics in Twentieth Century Urbanism*, Edinburgh, UK: Edinburgh University Press.

Raunig, G. (2007) *Art and Revolution: Transversal Activism in the Long Twentieth Century*. Derieg, A. (trans.), Los Angeles, CA: Semiotext(e).

Rancière, J. (1998) 'The cause of the other'. In *Parallax*, Vol. 4, No. 2, 25–32, I.

Rancière, J. (2010) *The Emancipated Spectator*, London: Verso.

Raven, A. (1989) *Art in the Public Interest*, Ann Arbour, MI: UMI Research Press.

Ross, K. (2002) *May '68 And Its Afterlives*, Chicago, IL: University of Chicago Press.

Sadler, S. (1999) *The Situationist City*, London: MIT Press.

Schechner, R. (1993) *The Future of Ritual: Writings on Culture and Performance*, London: Routledge.

Scott, E. (2016) 'Archives of present future: On climate change and representational breakdown'. In Graham, J., *Climates: Architecture and the Planetary Imaginary*, Zurich, Switzerland: Lars Muller Publishers, 130–141.

Situationist International (1960) 'The Minute of Truth'. *Internationale Situationniste #5*, December 1960. (trans) *Not Bored* (2007). Available at www.notbored.org/minute-of-truth.html.

Situationist International (1968) 'Report on the Occupation of the Sorbonne'. Council for Maintaining the Occupations, Paris, 19 May 1968. In Knabb, K. (ed.) *Situationist International Anthology* (Fourth Edition), Berkley, CA: Bureau of Public Secrets, 438–441.

Standing, G. (2014) *The Precariat: The New Dangerous Class*, London: Bloomsbury.

Virilio, P (2012) *Administration of Fear*, Cambridge, MA: MIT Press Semiotext(e).

Warner, M. (2002) 'Publics and Counterpublics' (abbreviate version). *Quarterly Journal of Speech*. New York Vol. 88, No. 4, 413–425.

10 New genre public commission?

The subversive dimension of public art in post-Fordist capitalism

Thierry Maeder, Mischa Piraud and Luca Pattaroni

Translation: Jessica Strelec

Long dominated by classical sculpture, art in the public space in Europe mainly played a commemorative role until the mid-twentieth century. The construction of allegorical monuments or monuments depicting incarnations of social order (political and military leaders, scholars or intellectuals) served as a staging of power, ultimately designed to justify a certain organization of the world (Ruby, 1998), and to legitimize States' actions while strengthening the collective identities of nations (Zask, 2013). Artistic production as a mode of subverting forms of established power did not exist *stricto sensu*, as public art was meant to be a reflection of it – although certain forms of critique or satire might have existed within commissioned art. Lefebvre likewise reminds us that all monuments, landscapes and spatial arrangements are the products of a ruling class, and therefore of power (Lefebvre, 1974). The late nineteenth century (and twentieth century to an even greater extent) nevertheless saw the emergence of more critical art forms – particularly as regards the conditions of its production and the challenging of its own codes. The result was the founding of various avant-garde movements, and later – often tied to revolutionary movements – of art that explicitly critiqued or more fundamentally subverted[1] society and its political and social organization. These works, which reflected art's subversive nature, are inherently dissentive and undermine capitalist, bourgeois social and political order in various ways (Kaprow, 1967).

This quest for subversiveness in art has created a paradox for public art commissions. To what extent is it possible for states to commission artistic interventions designed to upset the established order in public spaces?

In this chapter, we will defend the argument that this paradox has been partly attenuated by the historical integration of the subversive dimension of counterculture within the contemporary public commission process, and the modes of production of urban order more broadly. In other words, what was once subversive in terms of artistic intervention is now a common aspect of the interplay between art and urban development. Thus, for example, the *temporary* dimension of artistic intervention – once the crux of the subversive tactics of happenings – has become typical in the utilization of artistic events in urban marketing strategies. It is not that art has lost its characteristics *per se* – especially the ones that were the attributes of its 'radicality' – but rather that it has lost its power to upset the established order, that is, its subversiveness. Radicality must be understood as the relational attributes of art, one that is precisely manifested through its ability to subvert the established order. Such a relational approach calls, therefore, for a combined analysis of the evolution of art theory and practices and of the context of production and reception of artistic interventions.

In this perspective, we set forth to analyse the process of the relative loss of the subversive power of European and North American counterculture art in three steps. First, we will step back and look at the subversive potential of art as one possible outcome of its more fundamental political power. Next, we will analyse what exactly constituted the subversive dimension of the artistic interventions of the European counterculture of the 1960s and 1970s. Finally, we will show how those vectors of subversion have been contained by the new order of public commission and its broader relationship to the contemporary, post-Fordist, capitalist city. This is not to say that art has lost all its subversive potential or monumental functions, but simply that its ability to question the established order is no longer as apparent as in the 1960s and 1970s. As we will argue, public art finds its critical edge in the revamped relationship between art, politics and capitalism.

The politics of art

Underlying the question of art's subversive potential is broader thinking with regard to the art's role in the social and political realms, which in turn questions its role in the production of symbols and legitimation of forms of government.

For Eve Chiapello (and Luc Boltanski), the subversive potential of art is – or rather is *part of* – the *libertarian* dimension of the social struggles of the nineteenth and twentieth centuries that then combined with a *social* dimension ('social critique'), which they interpret as the struggle for equality (Chiapello and Boltanski, 1999). Heuristic though it may be, this approach separates politics (the desire for emancipation and democracy) and the social dimension (the desire for social justice and class struggles) somewhat too artificially. Hannah Arendt also seems to do this in *On Revolution*.[2] Art is limited neither to its discursive power nor to the commentary it makes on the world, which do not take into account the performative, pre-figurative and diagrammatic dimensions of what art does and makes distinctions that, historically, do violence to events, especially those of the Parisian events of May 1968 (Lazzarato, 2014).

To forge a more comprehensive understanding of art's political dimension (and, as such, its subversive potential), one can turn to authors like Gilles Deleuze and Félix Guattari, or more recently, Jacques Rancière. For Deleuze and Guattari, the subversive nature of art goes far beyond mere discourse or a shift of perspective (i.e. seeing the world differently). According to them, art moves lines,[3] and is therefore representative of the world to come (Deleuze and Guattari, 2005: 176, 183).

In a somewhat similar perspective, Rancière speaks of the 'politics of literature' in describing literature's purpose and the lines it shifts. However, we feel it would be more accurate to speak of 'politics of art' in general, to better reflect the power of art (Rancière, 2009: 591–606; Nicolas-Le Strat, 2015: 36). In *The Distribution of the Sensible* (2000), he evokes an aesthetic that is the basis for all political organization in the world, in that it defines what is – figuratively speaking – visible and audible. It is this *sharing*, as distribution and re-distribution of the sensible, that determines who can be seen and heard, who can see and tell, and, consequently, who organizes the space each person occupies in participation in the 'Common'.

With Rancière's approach, *all* artistic practice is highly political in its relationship to the world, and gives rise to its techniques and forms, regardless of the message that it does or does not purport to convey. For Rancière, for example, it is no coincidence that the rise of the revolutionary movements of the turn of the twentieth century coincided with the emergence of new art forms that blurred the boundaries between mediums. This redefinition of the relationship between observer and the observed led to a reformulating of the world's political organization.

It is initially in the interface created between different 'mediums' – in the connections forged between poems and their typography or their illustrations, between the theatre and its set designers or poster designers, between decorative objects and poems – that this 'newness' is formed that links the artist who abolishes figurative representation to the revolutionary who invents a new form of life.

(Rancière, 2000 [2006]:16)

Within this broader understanding of the founding role of artistic expression, art (and its *encaissement*[4] in the form of 'culture') can no longer be treated as a mere reflection of processes of production as it was in certain obtuse forms of Marxist orthodoxy. In this perspective, Gramsci marked an important milestone in overcoming this narrow conception of culture as a superstructure. For Raymond Williams:

It is Gramsci's great contribution to have emphasized hegemony, and also to have understood it at a depth which is, I think, rare. For hegemony supposes the existence of something which is truly total, which is not merely secondary or superstructural, like the weak sense of ideology, but which is lived at such a depth, which saturates the society to such an extent, and which, as Gramsci put it, even constitutes the substance and limit of common sense for most people under its sway, that it corresponds to the reality of social experience very much more clearly than any notions derived from the formula of base and superstructure.

(Williams, in Durham and Kellner, 2006: 134–135)

Art – and more generally all that participates in the (re)configuring of the sensible and intelligible framework of our experience – thus fully contributes to the mode of production. Yet, while we started our reflection with the subversive potential of art, it is worth noting that art, as part of the *infrastructure,* can function either as a *shift* or as a *reinforcing* of the boundaries of a given political order. In any case, to have subversive potential, art's constitutive power to act and have an impact on the world must be recognized.

The hegemonic power Gramsci describes, therefore, is built on a broad basis in which culture is an important pillar in the reproduction of social relations well beyond the symbolic level. If capitalism is upheld by culture – by building an *ethos* that hinders the moral and intellectual empowerment of the masses (Tosel, 2005) – it is also *by* and *through* art that the revolution must occur, through the building of an organic culture that breaks with dominant ideology and redefines common sense and our resulting relationship to the world. For Gramsci, artists – like intellectuals and journalists – are instigators of a cultural order that will or will not lead to emancipation.

Rancière suggests that this organic culture plays out, for example, in the reinvention of typographies, like in the *Arts and Crafts* movement,[5] or in the shift from written to theatrical expression. The politics of art thus goes far beyond the merely symbolic content of art, and rather lies in its ability to contribute to the production of actual counter-hegemonies.

As Chantal Mouffe (2009) explained, the politics of art play out on a number of levels. Borrowing from Gramsci's concept of 'cultural hegemony' – as a participation of culture in the process of production more than a conception of culture as a 'superstructure' – Mouffe, in her theory of agonistic democracy, offers a unique perspective on the role of art and artists in the construction of political confrontation. For Mouffe, political arenas are, in fact, *made* of this confrontation (via democratic tools) between non-reconcilable, hegemonic political projects. These hegemonies, once established, rely at the same time on a system of practices and the production of symbols and identities to legitimize and sustain themselves. Capitalism, for example, has instilled the identity of the consumer – particularly through advertising – to the point that modes of consumption have become core elements in the creation of subjectivities. So art, as an important creator of symbols, contributes to the founding of an *ethos* and is on the frontline of the reproduction – or challenging – of hegemony. That is why, for Mouffe, the political dimension of art, whatever form it takes, is undeniable:

> In that context, artistic and cultural practices are absolutely central as one of the levels where identifications and forms of identity are constituted. One cannot make a distinction between political art and non-political art, because every form of artistic practice either contributes to the reproduction of the given common sense – and in that sense is political – or contributes to the deconstruction or critique of it. Every form of art has a political dimension.
>
> (Mouffe *et al.*, 2001: 98)

Specifying this latter point in later texts (2007, 2014), Mouffe identifies four ways of intervention for critical artistic production. The first is direct involvement in political debate, which could be called commitment through 'message'; the second is a form of involvement through the exploration exclusion, marginalization and victimization phenomena; the third is self-reflection (by artists themselves) regarding the political conditions of their production; and finally, involvement through experimentation and the creation of utopias that oppose the values of the capitalist *ethos*. It is ultimately in this last form that art can best deploy its subversive potential in deconstructing common sense. In this respect, Mouffe again joins Gramsci; her artist – the initiator of new common sense – is similar to the Gramscian concept of the organic intellectual who lays the symbolic and semiotic bases of revolution. Mouffe maintains the essentially political dimension of art. As she well shows, this political nature cannot simply be reduced to forms of 'socially-engaged art', which is political in the very message it conveys. This form of (public) involvement of art, which functions on a meaningful level, couples with what it does semiotically above and beyond 'simple' meaning. We can say here that art functions 'diagrammatically' – in Deleuze and Guattari's sense of the word – that is, not (only) by representing the world differently but by moving its lines of force.

In an earlier work (Piraud and Pattaroni, 2016), we identified three levels of politics of art (i.e. levels on which art can make a *difference* or contribute to the reproduction of a *hegemony*) which, though based on different theoretical foundations, in a certain regard join the four levels of which Mouffe (2007, 2014) speaks. The first level acts as a process of *signification*, with a long tradition of political art that focuses on the *message* expressed and its various forms. Based on interpretative elements, this regime functions via the *semiotics* of its content. We have called the second level of overlap the *diagrammatic* regime, borrowing Guattari's term. This regime performs the function of a signifying and a subjective power of art. 'The work of the diagram [...] tends to the "capture of forces", to making "insensible forces sensible"' (Alliez, 2012: 10); put differently, how art interacts with the world and influences identities and subjectivities. On the third level, art was used in politics via its *processes of production* – how art functions in the market system, the spaces it occupies, its temporality and distribution networks, and so on, all of which are examples of what determines contemporary artistic production and the function it serves.

What should be remembered, from what has been said thus far, is that the political implications of art cannot be reduced to its critical or subversive dimensions. Rather, these implications play out in the way that art contributes to creating a shared reality. Thus, insofar as it figures and prefigures the world through the distribution of the sensible and the creation of subjectivities, art is *fundamentally* political. Thus, as it produces or reproduces a common sense that either moves or upholds the lines of force of an established order, it takes part in the politics. That is to say, the political nature of art is not limited to subversion but rather is the intrinsic nature of art.

Nevertheless, one can argue that the artistic movements at the root of the European and North American counterculture of the 1960s and 1970s – influenced by Adorno among others – associated the political dimension of art with its subversive dimension. We must now seek to better grasp wherein lay their subversive potential relative to the then prevailing and hegemonic symbolic and sensitive established order – that is, in Rancière's terms the 'distribution of the sensible' of the times.

Constituents of the subversive potential

The artistic movements or groups (e.g. The Situationists, The Living Theatre, Fluxus or their predecessor, Dada[6]) that influenced or directly participated in the creation of what has been be broadly referred – mainly in the European and North American context – as the 'counterculture' (Williams, 2006), shared certain strategies for imbuing their artistic intervention with a disruptive effect. Among others, one thinks of the intertwining of playfulness and spontaneity of their interventions (e.g. The Situationists' '*dérive*' method; Debord, 1958), the blurring of the social and visual borders between the public and private spheres, artist and audience (Schechner, 1965), or between life and art more generally. These strategies disturbed and disrupted the hegemonic distribution of the sensible of the Fordist system and its urban correlate – the functionalist city – based on strict and linear borders between the public and private spheres, between work, domestic life and leisure, and adherence to rationalist planning methods.

It is impossible to analyse all of these tactics, which likewise permeated and transformed the repertoire of political action (Cogato *et al.*, 2013), in a single chapter. Rather, we will focus on one of the key constituents of the counterculture's subversive powers: the systematic recourse to temporariness and ephemerality.

Indeed, at the roots of artistic protest movements claiming to be *anti-art*, like Dada or the Fluxus group, one finds a systematic recourse to performance and happenings based on a principle of temporariness.[7] This mode of artistic production was meant to be subversive as it presented various sources of disruption. First, it created a rupture with the world's long-term temporal organization – another central element of the distribution of the sensible – that is one of the conquests of capitalism (Thompson, 1967). Second, ephemerality also objects to what was one of the first justifications for art in the public space, namely the commemoration and expression of established power, which, by analogy, necessitate sustainability. Finally, the subversive power of ephemerality also resides in the fact that it allows for an escape from the canonical framework of the production system of artistic value. In *Loft Living*, Sharon Zukin shows how, in the 1960s, some artists, in rebellion against galleries and the art market, began to create works that could not be owned – notably monumental works and land art (Zukin, 1982). As they were non-transportable, these works required movement on the part of the audience to view them and made inclusion in any collection impossible. This refusal to be 'captured' in a collection also sometimes came with a rejection of the label 'monument'. In this regard, Gordon Matta-Clark's cut-outs,[8] for example, act as 'non-uments' (Kirschner, 1985).

Similar thinking is behind the idea of performances and ephemeral installations. By becoming a 'moment' rather than an object – by separating from its own materiality – art escapes the logic of accumulation that is the basis of capitalist functioning. And this is deeply subversive in relation to an art market that works by transforming the piece of art into a tradable object with a specific value.

Ephemeral approaches played an important role in the evolution of public art. Indeed, in response to 'official', 'imperious' public art, which was seen as a vehicle of social exclusion, the 1960s onwards saw the emergence of new forms of anti-establishment artistic expression, often allied with urban struggle movements seeking to draw attention to the struggle for control of the urban environment and its aesthetics (Sharp, Pollock and Paddison, 2005). Termed 'new genre public art' (Lacy, 1994) by artist Suzanne Lacy, these movements – which were critical of the role of artistic actors in exclusive urban development that promoted the accumulation of capital in the hands of ruling classes – led to artistic interventions in public spaces aimed at putting art back into the democratic debate by challenging underlying forms of domination.

Ephemerality, thus, became one of the tools, along with festive events and techniques such as The Situationists' *'dérive'*, counterculture art used to denounce and subvert the process of commodification of the urban space and the resulting exclusion.

One can argue that the idea of the subversive dimension of ephemerality gained momentum, and was generalized with the shift toward something akin to a *post-modernity*, after the avant-gardes – what Guattari called the 'winter years' of the 1980s. Indeed, following Soviet sclerosis, when the alternative to state communism had ceased to be attractive, the revolutionary idea undoubtedly moved towards an *insurrectionary* target: permanent movement, and acceleration became a way of objecting to any form of established authority. This turning point is particularly palpable in Hakim Bey's famous *Temporary Autonomous Zone* (*TAZ*), originally published in 1985. For Bey, 'reaction' always follows revolution, 'like seasons in Hell', and, therefore, is no longer enviable since it inevitably ends in oppressive stabilization in an order that smothers potential subjectivity. ('The slogan "Revolution!" has mutated from tocsin to toxin, a malign pseudo-Gnostic fate-trap, a nightmare where no matter how we struggle we never escape that evil Aeon, that incubus the State, one State after another, every "heaven" ruled by yet one more evil angel' (Bey, 1997: 5).)

Thus, Bey gives an exemplary interpretation of the subversive and emancipatory power of ephemerality. With a wary eye towards any form of institution, Bey wrote of TAZ, '[it] temporarily occupies a territory in space, time or the imagination, and dissolves when identified' (ibid.: 5). In the tradition of 'pirate utopias' and 'cyberpunks', it is a permanent de-territorialization that acts as shared potential.

> Like festivals, uprisings cannot happen every day – otherwise they would not be "non-ordinary". But such moments of intensity give shape and meaning to the entirety of a life. The shaman returns – you can't stay up on the roof forever – but things have changed, shifts and integrations have occurred – a difference is made.
>
> (ibid.: 5)

In a way, *TAZ* therefore echoes *A Thousand Plateaus*, wherein Deleuze and Guattari developed a theory of nomadism as a war machine.[9] This idea of subversion through movement undoubtedly joins and expands the thinking of Peter Bürger (1974) and Jean Dubuffet (1986), for whom the avant-gardes and artists, respectively, had to emancipate themselves from the threat that the institutionalization of art poses, that is, the 'asphyxiating' power of Culture with a capital C.

Those tactics of the ephemeral were also linked with attempts to give power back to the audience, which are another source of subversion of academic art. The Dada movement already invited, in the 1920s, the public to participate in its artistic happenings (Bishop, 2006).

The commandeering of subversive art?

We will now analyse the apparent paradox between public commissions and subversive art in greater detail. As we have suggested, this paradox is only partially true, as many processes have occurred since the 1960s that profoundly change the subversive impact of counterculture art, and the impact of temporary tactics in particular. To summarize, ephemerality is less and less a vector of subversiveness in societies where the distribution of the sensible is based on mobility and events. As we will argue in our conclusion, it is now, on the contrary, *continuity* that has become subversive. Thus, while the critical and subversive potential of art and artists today is undebatable, the conditions and context of this critique and subversion are themselves questionable.

De facto, the contemporary paradox of the commandeering – by public or private actors – of artistic practices largely influenced by the avant-garde movement is multifaceted. Before focusing on how the subversive power of temporariness has been partially contained, it is important to briefly explore some of these different facets which, as we will see, centre around the various dimensions of the three levels of politics of art mentioned earlier.

A first, critical point is that of art's autonomy in a context where artists depend on grants and acquisitions from public institutions[10] to survive – a point raised by Rainer Rochlitz (1994). One may ask to what extent the states related actors – employees of cultural public offices or elected politicians who function as important patrons of artistic production – accept criticism, the questioning of order (of which it is the manifestation) and, more importantly, its own functioning. The 1960s saw the propagation of artistic interventions, the critical dimension of which was a constitutive part. This often played out in the first level of overlap between art and politics – the symbolic level of the relationship between the signifier and the signified – such as in cases where politically explicit content of a publicly commissioned exhibition is called into question.

In Switzerland, one of the most resounding examples of such a clash between political interests and artistic autonomy is what the press called the 'Hirschhorn case'. In 2004, the Swiss Cultural Centre in Paris hosted Thomas Hirschhorn's *Swiss-Swiss Democracy* exhibition. Hirschhorn, who is extremely critical of the Swiss democratic system (whereby, in 2003, an extreme right-wing leader was elected to the federal government), caused a significant media and political scandal, giving rise to a parliamentary debate after which Pro Helvetia, the State-run foundation responsible for the Swiss Cultural Centre, saw its budget slashed by one million francs (Dubey, 2006). In addition to the controversial content of the exhibition, the issue also laid in its *mode of production*, that is, the fact that the exhibition was presented in a federal institution and that this institution suppos-edly represents Switzerland's cultural production in a foreign country. One could speculate that such an exhibition in a private venue would not have ruffled so many feathers. Such cases are undoubtedly an exemplary illustration of the way, even if public funds are enabling artistic interventions, they also hold the many attachments through which the subversive potential of art is contained. Such con-troversies lay the foundations for necessary debate on the notions of freedom and artistic exception vis-à-vis (public or private) sponsors' power.

Moreover, as suggested in our brief analysis of the fundamental relationship between art and politics, there are many other ways – in addition to the critical judgment expressed through works – through which the former may take on a subversive role. To grasp them, it is particularly interesting to turn to the case of art in the public space. Indeed, it is exemplary of this complex encounter with the question of politics and criticism as it is inherently subjected to legal scrutiny and order. In this regard, three points can help us consider the possible weakening of the critical power of artistic intervention in the public space.

First and foremost is what might be considered the subjugation of cultural poli-cies to the production strategies of the city. In keeping with the thinking on the creative economy underlying the debate with regard to urban development, sup-port policies for artistic production and exhibition are no longer merely cultural tools but tools of economic promotion and marketing. Through the concept of the *Artistic Mode of Production*, Sharon Zukin showed how the increasing valuation of artistic production (and particularly its consumption) in America in the 1960s was commandeered by market forces to transform and redevelop urban centres (Zukin, 1982).

Since then, numerous works have explored the links between cultural policies and urban strategies (artistic support and housing (Deutsche and Ryan, 1984); museum policies (Evans, 2003; Maeder, 2015; Olu, 2008); and State-funded public art programs (Cameron and Coaffee, 2005; Hall and Robertson, 2001; Mathews, 2010)). The installation of works of art in the public space has been legitimized by discourse based on socio-spatial expectations (i.e. creating a con-temporary aesthetic, strengthening local identity, attracting investments, etc.) that were later subsumed into the 'Creative City' discourse (Florida, 2002; Landry, 1995). That is to say, since the late twentieth century, art in the public space is no longer justified simply by arguments about the meaning of works, but rather its externalities and potential effects on the urban fabric. This integration of pub-lic art into the logics of production of urban forms has been likened, by certain authors, to the instrumentalization of artists, to legitimize a kind of subordination and reconquering of the urban space by State and financial powers (Deutsche, 1992). In a certain respect, such artistic interventions function in the same way monuments commissioned to proclaim the glory of a king or the wealth of a city once did. This is characteristic of the new functioning of cognitive capitalism and the process of spatial production it fosters, in order to reinvent monumentality, artists' involvement in improvement and enhancement policies.

Second, public art's high exposure to the critical judgment of the media and the public – not to mention fear of being a source of controversy for those involved in public commissions – tend to orient production towards more consensual forms. The rejection of contemporary art by public opinion has been the subject of numerous works, especially in France subsequent to (and undoubtedly given the failure of) policies to democratize access to contemporary art (Heinich, 2000; Michaud, 2011; Ruby, 2002). In an article published in 1989, Patricia Phillips denounced the standardization and bureaucratization of public commission programs leading, she claimed, to the smothering of more anti-establishment forms of expression. For Phillips, the complexity of selection processes, the increasing inclusion of the uninitiated (to make art 'more accessible' to society) and constant fear of scandal have led to a mentality of minimum risk, and even self-censorship, in anticipation of negative reactions and criticism of work considered too subversive or too provocative:

> So every possible – and ludicrous – objection is raised at the early stages of the artist selection and proposal process, to anticipate and fend off any possible community disfavor. With programs dependent on such tightly woven sieves, it's not surprising that plenty of hefty, powerful projects don't make their way through.
>
> (Phillips, 2004 [1989]: 195)

Steven Dubin studying the Artist-in-Residence program in Chicago in the 1980s, demonstrated how the meeting of diverging interests (of artists, curators and politicians) produces a specific form of culture and social organization from which emerge safe, uncontroversial works that are easily accessible to the public (Dubin, 1985, 1986).

Public art and containment

Alongside these two processes already broadly addressed in the literature, one can identify a third, less explored one that touches more directly on the crux of subversive works as they have developed in the second half of the twentieth century. Indeed, today we are witnessing – at least in European and North American post-industrialized democratic countries – the trivialization of artistic forms, both in their production methods and the sensible registers they produce, that were once the prerogative of alternative or protest movements by hegemonic groups and institutions. This holds true, for instance, for modes of expression like tagging and, more generally, the lessening of the *semiotics* (and 'aesthetics') of the counterculture for the commercial motives of contemporary design, thus dulling the subversive impact resulting from the use of alternative production methods (self-management, illegality, etc.) (Carmo *et al.*, 2014). More subtly still, it is also the case, as we would like to show in this chapter, of the banalization of temporary artistic intervention formats. Whereas once a tool of subversion, ephemeral has, in recent years, become one of the mechanisms for incorporating art in the city's systems of market production. Through festivals and happenings, ephemerality has become part of the contemporary broadening of the legitimate and market-oriented modalities of public art. To understand this shift, we must return to the very evolution of capitalist forms and the way it contributed to the generalization and legitimization of an order based on mobility (and the ephemerality of projects).

The development of post-Fordist capitalism – what Guattari called 'integrated world capitalism' – is the result of the incorporating of a desire for autonomy, decentralized management and, in a way, the modes of artistic production that characterized the critical movements of the 1960s and 1970s, into a market system (Chiapello and Boltanski, 1999). As works that thematise 'cognitive capitalism' show, *immaterial productions,* which represent an increasingly significant portion of production in general, are much more efficient when produced in non-hierarchical networks, which allows for greater use of general intellect (Moulier-Boutang 2007; Negri, 1991; Hardt and Negri, 2000, 2004). In other words, we again find here the moving, nomadic, ephemeral act – conceptualized in *A Thousand Plateaus* and *TAZ* – as the carrier of the radical production of a 'Common' in movement. Yet, this permanent deterritorialization has proven useful and extremely effective in liberating value added, and thus has been incorporated into the managerial lexicon (Chiapello and Boltanski, 1999). Some see this (relative) horizontalization of productive and reproductive relationships as a step towards emancipation, the beginning of the end of oppressive hierarchies, or even 'communism in capitalism' (without anyone daring to deny new specific forms of exploitation). Others, however, see it as a shameless 'distorting' of the social struggles of the 1960s and 1970s.

In such situations, where mobility has been placed at the core of the capitalist production of both goods and urban development, the danger of artistic movements claiming ephemerality and interstitial actions is their tendency to make order acceptable, a kind of carnival or valve effect that turns ambiguity into criticism in action. Hence something happens, a kind of integrating of the critique that becomes acceptable due to its ephemeral nature.

This process of incorporation leads to a partial neutralization of the subversive potential of artistic production (Mouffe, 2008). In a previous paper (Carmo *et al.*, 2014), we identified the phenomenon whereby the broad dissemination of formal registers produced through alternative or oppositional practices freed such forms from their original relationship of meaning by turning them into 'floating signifiers', in the words of Levi-Strauss (2003: xlix), and thus stealing their critical potential.

The same type of process is at work in art in the public space. As already said, in recent decades, this type of intervention has gradually become part of public commission programs. These movements, once dissentient, critical and anti-institutional, saw their methods and their militant, socially-engaged semiotic systems (and aesthetic?) trivialized to the point that 'while in theory refractory, [they found] an institution eager to get on board, and that encouraged them through direct solicitation: production, artistic commissions and events' (Ardenne, 2009: 195). And so emerged a series of initiatives – mainly at the instigation of local authorities through public art programs and urban policies – aimed at getting art back into public hands. At the heart of these approaches was the idea of partici- pation (Hall and Robertson, 2001), which can take many forms – from genuine inclusion in creative processes (process-based or process-oriented approaches) to consideration of expectations in the public contracts. In reality, however, these approaches were more often than not devoid of any critical basis.

Their subversive *mode* has, to some degree, been incorporated into institutional public commissions through its system of symbols but, more importantly, through the filtering down of approaches that were once the prerogative of alternative spheres – such as the ephemeral and the participatory. While these approaches originally aimed to step outside the lines by upsetting conventions and uses of urban space (e.g. the urban drifting of the Stalker group, or Stanley Brouwn's series *This way Brouwn*[11]), their commandeering by the State seems to have had the opposite effect.

Through commandeering, ephemerality becomes a perfect tool for the 'eventi- fying' of cultural policies and programs for art in the public space, which are now designed to serve cities' economic development and help commodify urban space and time (Ambrosino, 2012; Chaudoir, 2007; Vivant, 2008). Here again, the sub- versive dimensions underlying the idea of ephemeral art have been neutralized. Conversely, temporary actions are increasingly seen as a way not to consolidate and fix political action; but to minimize potential controversies and act quickly to address specific, measurable policy objectives (e.g. improving neighbourhood image, awareness, mediation, social work, etc.) by 'governing through the objec- tive' (Thevenot, 2011, 2014).

Participation underwent a process that is quite similar to that which affects temporary art production, and made it drift away from the intentions of the pio- neers, like Dada or Happening artists, to blur the boundaries and the conventions of art (Schechner, 1965; Butt, 2001). Attempts to include the 'audience' – with the multiple realities this word may define (Zask, 2013) – in the production of works began to abound in the 1990s, presumably when works ceased to be the initiative of isolated artists, and instead became an integral component of cultural democ- ratization policies. In public commission procedures, the participatory argument is often justified by the desire to reduce the cognitive distance between creation and exhibition of works, to bring the artist closer to her or his audience. While the participation process might be an interesting vehicle for mediation, it is not without negative consequences.

Indeed, this approach wherein artistic production is more or less subjected to the likes and dislikes of the public, can be seen as a form of populism. By implic- itly comparing co-produced, accessible and aesthetically pleasing art with press- ing, elitist art that is deemed to 'uglify' the urban space, discourse extolling the inclusion of the public is akin to populism (Deutsche, 1992), becoming a tool of political strategies. Thus, in Geneva, it is interesting that the first demand request by the public for inclusion in the artistic selection process in the late 1970s came from Vigilance – a far right-wing party[12] – which proposed that the winners of public art competitions be chosen by popular vote. The intent here was clearly not emancipatory, but rather opposition to the city's artistic policy acquisition, with the argument of preserving the public interest. Note that between the 1960s and 1990s, a number of major controversies associated with the installation of public works took place in Geneva. In 1982, the same party mobilized the residents of Alfred-Bertrand Park to object to the installation of a monumental work given to the city by American artist Dennis Oppenheim.

More generally, we can defend the idea that the blurring strategies of the 1960s – where art and life were one – had a perverse effect, as the appraisal of commissioned artistic interventions has changed not only in terms of their artistic value, but also their compliance, with increasingly complex urban regulations (hygiene, security, mobility, etc.). For example, as Nathalie Heinich suggested, in considering the controversy surrounding Serra's *Titled Arc* and several other cases of criticism addressed to public artworks, the use of 'civic principle' to judge public art is to a certain extent more legitimate in the United States than in France (Heinich, 2000). This seems true as regards the contemporary commissioning of temporary artistic interventions. Thus, in a recent public event in Geneva, where various artists were invited to produce work on the relationship between the city and the countryside, a proposal to put cows in a suburban middle-class housing project was denied by Geneva's state health services. Because of their temporary nature and blurred boundaries – and the relative ambiguity of what art is and is *not* – the very essence of artistic interventions has been stifled by the need to fit into the larger system of urban regulation, as norms allow for no exceptions.

The advent of the temporary and participatory dimensions – and hence attention paid to public reactions – in artistic interventions add to complex process of containing their subversive potential. Our final example aptly illustrates how a counterculture artistic project (and ephemerality in particular) ends up becoming part of the contemporary production of the city, so much so that is does not even raise public controversy. The Geneva association Espace Temporaire (ET) is an interesting example of the ambiguous effects of counterculture discourse and tactics (and temporariness most notably), resulting in both a renewal of critique and a lessening of art's subversive potential. ET (described as 'a laboratory for art and collective action') clearly thrives on what could be called an alternative grammar, meaning a repertoire of strong political proposals for the three facets of the politics of art. The title is a more or less direct indication of *TAZ*'s influence in Geneva's alternative community (the text was circulated in squats and was often discussed and alluded to). ET draws from the lexicon of the interstitial – 'Off spaces are interstitial, off screen spaces still to be discovered, created or simply imagined' (Espace Temporaire) – and the ephemeral. The idea is clear:

> The temporary occupation of the urban space questions the notion of a "clean and ordered" city. It generates experimental situations with social space,

creating areas of mutual dialogue and sharing. Through various mechanisms and off-kilter initiatives, Espace Temporaire experiments with the nomadic and spontaneous, imposes an unexpected presence, builds relationships, leaves a trail, and disappears only to reappear elsewhere.

(Espace Temporaire)

Among ET's more institutionalized activities, one finds a kiosk in a tramway station that has been turned into a small art gallery—a showcase for temporary public art. The kiosk's location and temporary nature potentially allow for disruptive displays. Nonetheless, as the following example shows, its temporariness also allows for reactivity and appeasement when public controversy rears its head. In spring 2013, the gallery exhibited *Cheval de Bataille*, by Maya Bösch and Régis Golay, a stuffed horse suspended from two straps, its head hanging, frustrated by 'all forfeited battles – large and small – especially those that remain invisible, hidden in the streets, behind closed windows, in the hearts of humans' (Bösch and Golay, 2013). The horse, long forgotten in the warehouses of the *Comédie de Genève*, was acquired in 2005 for a staging of Shakespeare's *Richard III* (as 'both accessory and centrepiece'). The artwork is simple, but the idea is scathing and serves as reference to Richard III who, 'during the inexorable conquest of power, suffered absolute defeat'. The reactions of the public were violent, and resulted in a major media and political controversy, notably via readers' letters to local papers ('You artists deserve to be hanged instead of the horse'). The exhibition was quickly shut down by the curators themselves, due most likely to considerable pressure and fear of political repercussions against the gallery.[13] The complex artistic proposition, echoing among other things Maya Bösch's staging of *Richard III*, illustrates Phillips's idea that, to be successful, public art must not make waves. 'Battle Horse', however, with neither cartel nor comment, did not result in the work's success, but instead led to self-censorship by the curators.

Ephemeral interventions, therefore, tend to lose their 'guerrilla' edge, becoming part of an urban order that systematically oversees all the entities that exist within it and pawns in larger marketing strategies. Though the potentially subversive diagrammatic effects of temporary tactics partially lose their disruptive potential. The same is true for the subversive dimension of temporary and collective interventions, as they take place within a broader system of production based on their temporary and collective nature.

Conclusion

The question today is whether subsidized public art – being inherently political, like any art form – can allow itself to be subversive, not only in terms of the message it conveys, as Rochlitz (1994) seems to say, but also in terms of production and diagram, that is, not only by adopting specific forms or content with the aim of raising controversy, but by truly shifting our relationship to the world and (in the words of Rancière) redefining the lines of distribution of the sensible.

Subversion through ephemerality seems unstable in a context where the latter is but the tatters of the hegemony of cognitive capitalism. Yet, we must recognize the ambiguity that is critiquing a mode of production – and the suffering produced – with the mode itself: it is a difficult to attack 'flexible capitalism' using the mode of flexibility.[14] We must undoubtedly distinguish here between the ephemeral and the reversible. None of Gordon Matta-Clark's cuts are still visible (except on video); nor do any of the buildings exist anymore. His *Splitting*, as a radical critique of suburban housing, has disappeared. Similarly, artistic conventions never recovered from the assault of 'ready mades'. Presumably, the critical potential of ephemerality only functions when it is *irreversible*, in the turmoil of a world it provokes (by eliminating the seriousness and nature of conventions, building new social groups, revealing new subjectivities, etc.).

We feel it is important to keep in mind not so much the fact that art can or cannot be subversive – depending on whether it is commissioned or not – but rather that art's critical potential is prone to daunting complexification (complexification of the constituent relationships between its modes of production, authors, commissioners and audience). While public art can be subversive, this subversiveness lies not in its content, shape or the sensible methods of production, but in the world it creates. This includes the interference and movement, deterritorialization, the relationship it establishes with the world and the critical distance it allows us to adopt.

This leads us to posit more generally a relative autonomy between modes of production, on the one hand, and diagrammatics on the other. A work that has all the finery of subversion can actually contribute to the reproduction of a hegemony, while a work produced according to the canons of traditional can – through the signs and symbols it uses, subjectivities it creates, utopias it instils and conventions it denies – move the lines of force that make the world, notably by influencing the public's critical thinking, its understanding of space and social relations.

While some authors can assert a link between form and content, between 'the speakable and the visible' (Rancière, 2000 [2006]), or between technique and political tendency (Benjamin, 1998 [1966]), it seems that the shift towards cognitive capitalism results in a reshuffling of the cards, forcing us to allow for certain considerations regarding the notion of socially-engaged art. While any mode of artistic production can potentially be 'commandeered' by the hegemony, the only way for art to be subversive is for it to rely on something else, for example, a shift in perspective, a germinal idea that the world as we perceive it and the common sense that structures it are but political constructions.

Acknowledgments

We would like to thank the Swiss National Science Foundation who funded part of the researches presented here.

Notes

1   We use the concept of 'subversion' to address a broad range of ways an established order is questioned and upset. Explicit critique, working mainly on a symbolic or semantic level, is a constitutive part of the subversive power of art. But art's subversiveness comes also from its ability to question the established order on other levels, especially sensitive one.

2 On this topic, see Hardt and Negri's critique of Arendt (Hardt and Negri, 2004: 102).
3 For Deleuze and Guattari's, the world, individuals and groups are made of lines of force. Deleuze defines three types of lines: *molar lines*, which segment the world in rigid binary machines (male-female, child-adult, black-white, etc.); *molecular lines*, which disturb the symmetry created by molar lines and may create new segments that do belong to the binary categories ('molecular sexuality which is no longer that of a man or a woman, molecular masses which no longer have the outline of class, molecular races like little lines which no longer respond to the great molar opposition' Deleuze and Parnet, 1987 [1977]: 131); and *lines of flight*, which are the lines of deterritorialization and destratification: '[they] create an irreversible aspiration to new spaces of liberty' (Guattari, in Sasso and Villani, 2003: 210).
4 Johan Stavo-Debauge, reflecting on the experience of public problems, and notably a 'series of problems that are little considered, even though they are indicated by several expressions of ordinary language', proposes an interesting discussion on the French notion of *encaissement*. This idea 'mainly helps to voice the exercise of a certain violence' (to take a hit), but also packaging (i.e. putting something in a crate and thus containing it in a well-defined place), which can be extended more or less directly to profitability ('collecting funds'). We do not attempt to translate the French word, as it contains all three, interconnected meanings (Stavo-Debauge, 2012). In this perspective, culture can be seen as a process wherein the disruptive dimension of art is contained, allowing it to be integrated into various public policies and a profitable market.
5 The arts and crafts movement emerged in Britain at the end of the nineteenth century. It advocated the infiltration of fine art aesthetics and traditional skills into the production of everyday life objects.
6 Dada was a European artistic and intellectual movement of the avant-garde. It was established in Zürich (Switzerland) in 1916. Pacifist and anti-bourgeois, the members of Dada often used humour in their works, as well as new techniques of production (such as demonstrations and public gatherings).
7 For more on this, see also Sharp, Pollock, and Paddison (2005) and Lacy (1994) on the origins of grassroots artistic movements that use uncommissioned, temporary artistic interventions in the public space.
8 From the early 1970s to his death in 1978, the American artist Gordon Matta-Clark performed several cuts in abandoned buildings that were about to be demolished.
9 'The concept of psychic nomadism (or "rootless cosmopolitanism", as we call it in jest) is vital in the creation of the TAZ. Deleuze and Guattari discuss certain aspects of this phenomenon in *Nomadology and the War Machine*, by Lyotard in *Driftworks* and by various authors in the "Oasis" issue of the review *Semiotext(e)*. We prefer the term "psychic nomadism" to "urban nomadism", "nomadology", "driftwork", etc., with the simple aim of linking all these concepts into a single, fuzzy group to study in light of the emergence of the TAZ' (Bey, 1997: 9).
10 Heritage collections, regional/municipal contemporary art funds, etc.
11 In this work, Stanley Brouwn asks passers-by to draw a map to get to a point in the city of their choosing on a piece of paper, which he then attempts to reach based on the drawing.
12 Vigilance was active in Geneva from 1965–1991. It consistently opposed art projects in the public space in the city.
13 Indeed, a few months after the exhibition a city councillor asked to reconsider the lease between the city and the gallery and to transform the edifice in a café, his demand was notably motivated by the content of the past exhibitions.
14 For Hardt and Negri: 'What may have been most valuable in the experience of the White Overalls was that they managed to create a form of expression for the new forms of labor – their network organization, their spatial mobility, and temporal flexibility – and organize them as a coherent political force against the new global system of power. Without this indeed there can be no political organization of the proletariat today' (Hardt and Negri, 2004: 267).

References

Alliez, E. (2012) 'Diagrammatic Agency Versus Aesthetic Regime of Contemporary Art: Ernesto Neto's Anti-Leviathan'. *Deleuze Studies*, 6 (1), 6–26.

Ambrosino, C. (2012) 'Ces esthétiques qui fabriquent la ville'. In Terrin, J. J. (ed.), *La Ville des Créateurs: Berlin, Birmingham, Lausanne, Lyon, Montpellier, Montréal, Nantes*, Marseille: Parenthèses, 180–199.

Ardenne, P. (2009) *Un Art Contextuel: Création Artistique en Milieu Urbain, en Situation, d'Intervention, de Participation*, Paris, France: Flammarion.

Benjamin, W. (1998 [1966]), *Understanding Brecht*, London: Verso.

Bey, H. (1997) *TAZ: zone autonome temporaire*, Paris: L'Eclat.

Bishop, C. (2006) *Participation*, Cambridge, MA: MIT Press.

Bösch, M. and Golay, R. (2013), Le Cheval de Bataille. ZABRISKIE POINT [online]. Available from www.zabriskiepoint.ch/?page_id=1103 (Accessed 3 February 2016).

Bürger, P. (2013) *Théorie de l'Avant-Garde*, Paris: Questions théoriques.

Bürger, P. (1974) *Theorie der Avantgarde*, Frankfurt: Suhrkamp.

Butt, G. (2001) 'Happenings in History, or, the Epistemology of the Memoir'. *Oxford Art Journal*, 24 (2), 115–126.

Cameron, S., and Coaffee, J. (2005) 'Art, Gentrification and Regeneration: From Artist as Pioneer to Public Arts'. *European Journal of Housing Policy*, 5 (1), 39–58.

Carmo, L., Pattaroni, L., Piraud, M., and Pedrazzini, Y. (2014) 'Creativity without critique', paper presented at the Lisbon Street Art Urban Creativity International Conference, Lisbon, July.

Chaudoir, P. (2007) 'La ville événementielle: temps de l'éphémère et espace festif'. *Géocarrefour*, 82 (3).

Chiapello, E. and Boltanski, L. (1999) *Le Nouvel Esprit du Capitalisme*, Paris: Gallimard.

Cogato Lanza, E., Pattaroni, L., Piraud, M. and Tirone, B. (2013) *De la Différence Urbaine: Le quartier des Grottes/Genève*. Genève, Switzerland: MétisPresse.

Debord, G. (1958). Théorie de la dérive. *Internationale Situationniste*, 2, 19–23.

Deleuze, G. and Guattari, F. (2005) *Qu'est-ce que la Philosophie?* Paris: Ed. de Minuit.

Deleuze, G. and Parnet, C. (1987) *Dialogues*, New York: Columbia University Press.

Deutsche, R. (1992) 'Art and Public Space: Questions of Democracy'. *Social Text*, (33), 34–53.

Deutsche, R., and Ryan, C. G. (1984) 'The Fine Art of Gentrification'. *October*, 31 (winter), 91–111.

Dubey, M. (2006), 'L'affaire hirschhorn: De l'etat mécène à l'etat architecte' Masters thesis, University of Geneva.

Dubin, S. C. (1985) 'The politics of public art', *Urban Life*, 14 (3), 274–299.

Dubin, S. C. (1986) 'Artistic Production and Social Control', *Social Forces*, 64 (3), 667–688.

Dubuffet, J. (1986) *Asphyxiante culture*. Paris, France: Ed. de Minuit.

Durham, M. G. and Kellner D. M. (2006) *Media and Cultural Studies: Keyworks*, Hoboken, NJ: Wiley.

Espace Temporaire (n.d.) *Off Spaces/Espace Temporaire* [online]. Available from www.espacetemporaire.com/off-spaces/ (Accessed 3 February 2016).

Evans, G. (2003) 'Hard-branding the cultural city – From prado to Prada'. *International Journal of Urban and Regional Research*, 27 (2), 417–440.

Florida, R. (2002) *The Rise of the Creative Class: And How it's Transforming Work, Leisure, Community and Everyday Life*, New York, NY: Basic Books.

Hall, T. and Robertson, I. (2001) 'Public art and urban regeneration: Advocacy, claims and critical debates'. *Landscape Research*, 26 (1), 5–26.

Hardt, M. and Negri, A. (2000), *Empire*. Paris, France: Exils.

Hardt, M. and Negri, A. (2004), *Multitude: Guerre et Démocratie à l'Age de l'Empire*. Paris, France: La Découverte.

Heinich, N. (2000) 'From rejection of contemporary art to culture war'. In Lamont, M. and Thévenot, L. (eds.), *Rethinking Comparative Cultural Sociology*, Cambridge, UK: Cambridge University Press, 170–209.

Kaprow, A. (1966) *Assemblage, Environments and Happenings*. New York: H. N. Abrams.

Kishnei, J. R. (1905). Non Umento. *Artforum*, 24 (2), 102–108

Lacy, S. (1994) *Mapping the Terrain: New Genre Public Art*, Seattle, WA: Bay Press.

Landry, C. (1995) *The Creative City: A Toolkit for Urban Innovators*, London: Earthscan.

Lazzarato, M. (2014) *Marcel Duchamp et le Refus du Travail*, Paris, France: Les Prairies Ordinaires.

Lefebvre, H. (1974) *La Production de l'Espace*, Paris, France: Anthropos.

Lévi-Strauss, C. (2003) 'Introduction à l'oeuvre de Marcel Mauss'. In *Sociologie et Anthropologie*, Paris, France: PUF, IX–LII.

Maeder, T. (2015) 'Les grands équipements culturels et leurs enjeux: étude du pôle muséal de Lausanne'. *Cahiers De Recherche Urbaine*, 3, 1–119.

Mathews, V. (2010) 'Aestheticizing Space: Art, Gentrification and the City'. *Geography Compass*, 4 (6), 660–675.

Michaud, Y. (2011) *La Crise la l'Art Contemporain*, Paris: Presses Universitaires de France.

Moulier-Boutang, Y. (2007) *Le Capitalisme Cognitif: La Nouvelle Grande Transformation*, Paris, France: Editions Amsterdam.

Mouffe, C., Deutsche, R., Joseph, B. W. and Keenan, T. (2001) 'Every form of art has a political dimension'. *Grey Room*, 2(Winter), 98–125.

Mouffe, C. (2007) 'Artistic activism and agonistic spaces'. *Art and Research*, Vol. 1.

Mouffe, C. (2008) 'Art and Democracy'. *Open*, 14.

Mouffe, C. (2009) 'Democratic Politics in the Age of Post-Fordism'. *Open*, 16.

Mouffe, C. (2014) *Agonistique, Penser Politiquement le Monde*, Paris, France: Beaux-Arts de Paris.

Negri, T. (1991) 'Travail immatériel et subjectivité'. *Futur Antérieur*, 6.

Nicolas-Le Strat, P. (2015) 'The Ecosophic Conversion of Creative Practice or the Power of Indeterminacy / La conversion écosophique des pratiques de création ou la puissance de l'indétermination' Dare-Dare (ed.) *Dis/location 2: Projet d'Articulation Urbaine*, 25–36.

Olu, E. (2008) 'L'argument culturel du "touristique", l'argument touristique du culturel, symptômes de "la fin du muséal"', *Téoros Revue de Recherche en Tourisme*, 27(3), 9–17.

Phillips, P. (2004) 'Out of order: The public Art Machine'. In Miles, M., Hall, T. and Borden, I., *The City Cultures Reader*, London: Routledge, 190–196.

Piraud M.-S and Pattaroni L. (2016) 'Politiques des mondes de l'art: Signification, diagramme, production'. In Bellavance G., Fleury L. and Péquignot B., *Monde, Champ, Scène au Prisme des Réseaux*, Paris, France: L'Harmattan.

Rancière, J. (2000) *Le Partage du Sensible*, trans. Rockhill, G. (2006), *The Politics of Aesthetics*, London: Bloomsbury Academic.

Rancière, J. (2009) *Et Tant pis pour les Gens Fatigués. Entretiens*, Paris, France: Editions Amsterdam.

Rochlitz, R. (1994). *Subversion et subvention*. Paris: Editions Gallimard.

Ruby, C. (1998) 'Art en public ou art public?', *Le Débat*, 1 (98), 49–59.

Ruby, C. (2002) *Les Résistances à l'Art Contemporain*, Brussels: Editions Labor.

Sasso, R. and Villani, A. (2003) 'Le vocabulaire de Gilles Deleuze'. *Les Cahiers De Noesis*, 3.

Schechner, R. (1965) 'Happenings'. *The Tulane Drama Review*, 10 (2), 229–232.

Sharp, J., Pollock, V. and Paddison, R. (2005) 'Just art for a just city: Public art and social inclusion in urban regeneration'. *Urban Studies*, 42 (5), 1001–1023.

Stavo-Debauge, J. (2012). Des '"événements" difficiles à encaisser : Un pragmatisme pessimiste'. In Céfaï D. et Terzi C., *L'expérience des problèmes publics, Raisons pratiques, 22*. Paris: Editions de l'Ecole des Hautes Etudes en Sciences Sociales, 191–223.

Thévenot, L. (2011) 'Metamorphose des evaluations autorisées et de leurs critiques: L'autorite incontestable du gouvernement par l'objectif'. In de Larquier, G., Favereau, O. and Guirardello, O. (eds.), *Les Conventions dans l'Economie en Crise*, Paris, France: Éditions La Découverte.

Thévenot, L. (2014) 'Autorités à l'épreuve de la critique. Jusqu'aux oppressions du "gouvernement par l'objectif."'. In Frère, B. (ed.), *Le Tournant de la Théorie Critique*, Paris, France: Desclée de Brouwer, 216–235.

Thompson, E. P. (1967) 'Time, Work-Discipline, and Industrial Capitalism'. *Past and Present*, 38 (38), 56–97.

Tosel, A. (2005) 'La presse comme appareil d'hégémonie selon Gramsci'. *Quaderni*, 57 (1), 55–71.

Vivant, E. (2008) 'Les événements off: de la résistance à la mise en scène de la ville créative'. *Géocarrefour*, 82 (3).

Williams, R. (2006) 'Base and superstructure in Marxist cultural theory'. In Meenakshi Gigi Durham and Kellner (eds.), *Media and Cultural Studies: Keyworks*, Oxford, UK: Blackwell, pp. 130–143.

Zask, J. (2013) *Outdoor Art*, Paris, France: Éditions La Découverte.

Zukin, S. (1982) *Loft Living: Culture and Capital in Urban Change*, New Brunswick, NJ: Rutgers University Press.

11 Unpacking public places in Gothenburg – The Event City

Becoming a cannibal

Marika Hedemyr

I'm slowly dancing between the trees in the park. A brand new public park jammed in between the police station, the custody centre, the district court, and Ullevi Arena. Right in the heart of Gothenburg's 'Event District'. Dancing, measuring the distance and volume. Looking at the sky. Moving over the ground. Suddenly, in the corner of my eye, I see a police guard running towards me. His hand ready at the gun holster. I sense his adrenalin before I can see his face. "Is it your bag? What are you doing here?" I stop dancing and turn towards him. "I'm just dancing with the trees, as a research for a performance here," I reply. "You're fucking lucky I didn't press the bomb alert button. You can't leave your bag just like that, or move around like that." His hand is still at the gun when he speaks into his walkie-talkie: "Call off the emergency squad, it seems to be an artist."

(Hedemyr, 2014b)

In the public places of a gentrified, commercialized 'Event City', how can I respond to the socially urgent with an aesthetic imagination? As a local artist, how can I make a difference? These questions were my starting points when exploring the park. At first glance it was a quite uncomplicated place – a few trees, a grass lawn and open air. The city of Gothenburg in Sweden had recently bought the ground to turn it into a public park, but the small area was still under camera surveillance by at least three different guard teams in the surrounding buildings. Working on-site, as in the situation above, I experienced the contested borders of public/private, allowed/ forbidden and open/closed. It was a complicated intersection of conflicting perspectives and interests: private, public, commercial, social, political, law and order.

Public space in a city activates the complexity of humans being together in an urban environment. Public space is also a place where the city wants to present itself. Not necessarily reflecting how the city or situation is in reality, but how it would like to be perceived, and is thereby continually contesting the borders between reality and the image of the urban reality (Lefebvre, [1974] 1991; Massey, 2005).

The park and the buildings around it – the police station, the custody centre, the district court and Ullevi Arena – are located in Gothenburg's Event District (*Evenemangsstråket* in Swedish). It is a commercial area with large sport and music arenas, an exhibition and congress centre, an amusement park, hotels and museums. The Event District is branded and packaged as an area for events, exhibitions and entertainment, a meeting place for families and congress visitors alike. Located right in the city centre and within a fifteen-minute walking radius, the area has been a key element in branding Gothenburg as an 'event city'. That the area also is a residential area with schools, kindergartens, parks and businesses is not visible in the branding and marketing schemes.

The development of Gothenburg is similar to that of many former industrial mid-sized cities in northern Europe in its quest to transform itself from an industrial city into a university, information or creative city. The process has evident winners and losers from an economic and sociocultural point of view, with increasing spatial and socio-economic segregation. Several former ports, docks and industrial areas in the inner city are going through a transformation, which does not only involve a restructuring of the physical space, but also a branding process that contains changing narratives and the displacement of people. Investors and developers see the former industrial areas as 'abandoned' and 'empty' with the potential for building a new attractive city. However, these areas are rarely empty, even if no one lives there. On the contrary, they are often used by people who financially might have difficulties finding a place in or taking part in the increasingly upgraded event city.

Even the city's name has transformed into a brand. The Swedish name is *Göteborg* and spelled with two dots *above* the first o: the letter ö. In 2009 the two dots were pushed off the *o* and landed beside it, like a colon, and since then the citizens have lived in a brand – Go:teborg – that calls for tourists and businesses. Today Gothenburg has a great success in the visitor industry; and with the prefix 'Go:' the city is marketed and developed as a city of tourism, meetings and events.

In the creation of a successful and creative event city, perspectives tend to clash, creating a complex relationship with local artists and citizens. The branding process is recurrently appropriating the cultural capital of local initiatives, and the well-established collaborative spirit of the city is at times bordering on being corrupt (Lundström, 2012; Öberg, 2013). At the same time, it is also one of the most segregated cities in northern Europe. In the decades after World War II, Gothenburg was the fastest growing city in Sweden and its prosperity was guided by a distribution policy. The socio-economic gaps decreased and the socially mixed public schools contributed to social cohesion. The big shift in social development took place in the 1990s. The economic crises of the early 1990s in Sweden resulted in a restructuring of the Swedish economy, and a redrawing of the political landscape. Half a million jobs disappeared and Gothenburg was hit hard. The crises of the 1990s were resolved long ago, but the patterns of inequality and segregation that came forward still remain. Today, the city has alarming disparities in health, unemployment and child poverty between different neighbourhoods, and the difference between the rich and poor continues to increase (Sernhede, 2016).

My own experience of the city comes from being an artist based in Gothenburg – working both locally and internationally – and as a citizen, living in Gothenburg with my family. I work across choreography and public art, creating performative works for stage, art venues and public spaces, exploring issues of access, democracy and the relation between people and places. I have also been running a dance theatre company (Crowd Company 2000–2014) and a studio platform for theory and practice (Dansbyrån 2001–2016), both with their bases in Gothenburg.

Living in Gothenburg, it feels as if 'The Event City' is colonising and cannibalising its own locals in its pursuit of attracting visitors. Cannibalism is the practice of humans eating the flesh and internal organs of other human beings. Metaphorically, it is a critical way to digest the other. The idea is that by eating the other, you acquire some of the power and characteristics of the one you eat. The city is branded with arguments such as 'the city space as an arena' and 'event-loving locals' (Go:teborg, 2015), which might be great if you are a visitor, but is an abstract and almost violent description when you live in the city as one of these 'locals'. It creates a stifling cover over the complex reality of the city.

Since 2012, I have explored in which way one could critically and aesthetically approach the complexity of the successful event city, and create works that spur critical thinking around the issue. My experience is that if a work of art is criticizing a situation from a position of opposition, it is easily dismissed as 'merely art' or 'merely critique'. A contributing factor is the ambiguous perception of art's autonomy: in the romantic image of art as a field for reality-tests, a 'free zone' where otherwise unthinkable transformations are possible. In practice, the idea of the free zone has a preservative effect. Contemporary art is accepted as a free zone for radical ideas, system-critical works and political discussions – on condition that it stays within the boundaries of the art world (Jönsson, 2012). If art is to realize its full potential to both act critically and produce new modes of social encounters, it is not by operating from an *outside* position or a position of resistance by criticizing from a *distance*, but from a position *close* to the centre.

Since each location in the urban landscape holds competing narratives about its place, situation and relevance, I see a possibility to keep criticality in aesthetic works by unpacking public places and re-presenting them in a way that destabilizes the single story. An artwork can provide an entry point for experiencing the complexity of a place, or simply for experiencing the location differently (Doherty, 2015; Malzacher, 2014). It can reveal conditions of conflict by unpacking the particularities of both dominant and hidden stories of a place. Thereby, it has a potential to act critically.

On the following pages I will, from an artist/practitioner's perspective, share experiences and proposals that are exploring this strategy of staying at a critical proximity to the centre. The urban context is Gothenburg – The Event City. I will discuss my experience from *The Event Series* (2014–ongoing), a long-term art project I started in 2014, and one of the works in the series *Guide by Numbers* (2014), which I created in the park outside the police station. This work took the form of a guided tour and the topic of present and absent bodies and voices is a strong element. As an additional example, I will refer to the book *Den Urbana Fronten* (2015) by photographer Katarina Despotovic and writer/researcher Catharina Thörn.

Working with – becoming a cannibal

Cannibalism alone unites us. Socially. Economically. Philosophically.

(de Andrade, [1928] 1991)

My response as an artist to the issues of present and absent bodies in the public spaces of the event city, its rhetoric, and the particularities of the political inscribed in this local geography, was to become a cannibal myself. I aimed for the heart and aorta of the event city: The Gothenburg Event District.

Making this urban area my site and material, I started *The Event Series* in 2014. It is a long-term art project in gentrified and commercialized urban city centres that explores how it is possible to create and propose performative art and artistic practice that, through the body, talks about/to our society; reflects these urban environments in their complexity; invites people into critical reflexion, to have an artistic experience and exchange of ideas; creates both deconstruction and production on location; and moves in a landscape in between criticality and poetry. I create a series of site-specific works, and develop a discourse for the tacit knowledge and expertise involved in this practice that, by using the body, talks about/to and in public space about late modern life in the urban landscape.

The artistic strategy I explore is *working with*, close to the centre. Not against or in opposition, yet keeping a critical stance and keeping an artistic autonomy for the practice, my works and me. Cannibalism is an extended metaphor for this strategy, departing from *The Manifesto Antropófago* (*The Cannibalist Manifesto*) by the Brazilian poet Oswald de Andrade (1928/1991). The entire text of *The Cannibalist Manifesto* can be seen as a critique of colonial relations and proposes that cannibalism is like a critical way to digest the other. It is a creative way to incorporate what you admire in the other. It gives me a particular capacity to reinterpret the symbols and cultural codes. I eat them, chew them and then spit them out, combined in new ways. It provides a strategy to get to the very centre of a situation, in which I can operate in critical and transformative ways. Staying close to the centre and appropriating modes and expressions to play with them can create humorous and confusing outcomes. In the gap of confusion – is this real, or a joke; is it serious, or a change of direction? – there is a potential to reflect upon the current state of things, which can act as a catalyst for critical thinking.

Architect and urbanist Teddy Cruz suggests a similar position for art and artists. He advocates that artistic practice moves from a critical distance to a critical proximity, through which 'emerging activist practices seek, instead, for a project of *radical proximity* to the institutions, transforming them to produce new aesthetic categories that can problematize the relationship of the social, the political and the formal' (Cruz, 2012: 60–61, italics in original).

By cannibalising the material of a location and site, you can critically digest and then incorporate it in creative ways in your work.

So what do you eat?

What constitutes the 'material' of a location and site depends upon the artistic discipline and practice. Being an artist, working with artistic practice and outcomes can give access to information: rooms and contexts that otherwise might not be open to the public. One can operate between a state of embeddedness and critical distance, working with the place and situation as both an insider and outsider at the same time. But one can't escape the normative and stereotypical judgments one receives in public. Not for one's art, neither one's activities, nor one's appearance. As an artist, you are part of the material

> We are writing our own Walk of Fame with chalk on the pavement at The Avenue, the main boulevard in Gothenburg. We add names, phrases. It's an artistic experiment. Passers-by stop to have a look. Some comment on what we do. "Is this some kind of art?" This is the most common question I get. I am a white woman with Northern European appearance, with a chalk in my hand. My colleague is doing the same thing a couple of meters away from me. "Is this some kind of political protest?" This is the most common question he gets. He is a bearded man with Latin American appearance, with a chalk in his hand.
>
> (Hedemyr, 2014b)

When discussing the above situation and experiment afterwards with my colleague Hector Garcia Jorquera, we realized that, in Sweden, the image of the politically active person from Latin America is very strong. Therefore any activity out of the ordinary is first read as a political gesture. For me, a white Swedish woman, if I'm not acting out of mental illness, my activities are first read as an artistic gesture. These preconceptions and normative readings are all part of the material. It can be both an advantage and a disadvantage. They can be tiresome, these stereotypes and norms. One can't escape them, but one can cannibalise them. By shifting to work *with* these preconceptions with a cannibalistic approach, they can be used to provide precise observations when unpacking a public space, and an even sharper critique on the issue.

To get hold of the material and the fleeting stories of a place, I combine spending a lot of time at the location with talking to people from the local vicinity. It allows me to understand, expose and respond to the situation of a place. I work with an interdisciplinary choreographic practice; although my current works seldom look like dance. My artistic practice is located at the intersection of public-, performing- and site-specific art. The material I work with is found in the *site-specificity* of a location, which I perceive as what art historian Rosalyn Deutsche (1996, xi) describes as an interdisciplinary space and discourse that combines 'ideas about art, architecture, and urban design, on the one hand, with theories of the city, social space, and public space on the other'. The concept of site is not understood simply as a geographical location or an architectural setting, but includes a social situation: a network of social relations, stories, facts, events, present and absent bodies; and a choreographed manner in which bodies are encouraged to move, act, and think in relation to the site. I work with space as a product of interrelations, as a 'sphere of the possibility of the existence of multiplicity and the sense of contemporaneous plurality' recognizing that space is always under construction (Massey, 2005:9).

One way to get hold of the material of a place is simply to talk to people. I have come to call them 'experts', the people I meet who have knowledge or stories about a place. Often people surprise me by knowing a lot about a specific topic, detail, or aspect.

> She stretches her right arm out in front of her, raises her thumb, closes one eye, and is squinting, holding her thumb towards the top of the tree. "This tree here is 27 metres." She moves her hand and eye to another tree. "This one is 25 metres high, and that one's not more than 23. That one over there is 200 years old, and that one's probably not more than 150 years."
>
> (Hedemyr, 2014b)

I had met Eva-Maria Hellqvist, a tree expert working for the City of Gothenburg. It was part of my process of unpacking the park outside the police station, in what was to become the work *Guide by Numbers*. At a glance she could tell me the height and age of the trees, the depth and width of their root systems, and the names of every single plant in the park. By the scars on the bark, she could tell if a branch had been broken off naturally, or if it had been removed by a gardener. All this was valuable material for me. Another expert I met was Annelie from the ticket office at Ullevi Arena. She knew everything about the different setups for sport events and concerts, and how it influenced the number of the seats with restricted view.

Finding these experts is often about asking people at the location if they know someone who knows about a specific thing. One person leads to another, and suddenly one finds a person that is a master of his/her domain. Talking with locals has led me to a power shovel owner specialized in archaeological excavations, and to a driver who transports persons who are to be deported from the country. Another time, the questioning led me to a person whose position I didn't know. Before the policeman in charge at the reception of Gothenburg police station dialled the phone number, he hesitated for a moment and asked me: 'The person you are asking for is the head of the entire police force in West Sweden. Are you really sure it's him you're gonna talk to?'

Talking to these experts, without knowing what to look for, leads to magic moments when stumbling over some unexpected fact or anecdote that in some strange way connects to other information or practice. Sometimes the material itself guides by sheer coincidence, as in a section in *Guide by Numbers* where I included facts in the script:

> Those who are buried below ground here are relatively short, even for their time. Women are on average 1.60 metres tall, and men 1.68. The main reason

why they were short is thought to be the monotonous diet, which basically consisted of three options: porridge, gruel, and the drink *ölslupa*, which was a mixture of beer and gruel.

Today, the food here in town is better, luckily. For example, at the custody centre they provide three dietary options: normal diet, normal diet minus pork, and vegetarian diet. Down at the lunch restaurant here on the corner, they also have three diet options. They provide the options: 'Farm', which means meat; 'Sea', that is fish; and 'Meadow' – a vegetarian option.

(Hedemyr, 2014a)

What to do with the material from a site, and how to combine it into a work, depends on the material, the possibilities of the site, and the choices made by the specific artist. A key question in the process of *working with,* as a cannibal, is how to balance autonomy, responsibility and alliances with one's purpose and aim. Questions of how to select, compose and activate the site and material emerge. Especially in places where conflicting interests and perspectives are operating at the same time, and where some voices are very strong, others weak.

Chewing and re-presenting stories of public spaces

For an artwork to act as mediating object and a point of dialogue that can stand between the idea of the artist and the feeling and interpretations of the spectators, it needs to be a 'third thing' that both parts can refer to. Art historian and critic Claire Bishop, who develops an analysis of the relation between participation and spectacle, argues that to enlarge our capacity to imagine the world and our relations anew, art and the social need to be sustained in a continual tension. This 'requires a mediating third term—an object, image, story, film, even a spectacle—that permits this experience to have a purchase on the public imaginary' (Bishop, 2012: 45). To endure as a third object, whatever it is, it must keep an aesthetic integrity. Then it will have the capacity to open up to seeing things differently. If it loses this aesthetic integrity, it may lose its function as a mediating third object. Curator and researcher Claire Doherty stated that if it 'collapses' as a third object, it may offer pleasurable experiences and 'immediate gratification, but which do not generate new forms of critical dialogue or transformation' (Doherty, 2015: 16).

How does a work of art become a third thing with artistic integrity? There are no clear answers to that. The options are as many as there are artists. It depends on how material, form, aesthetics and actions are selected and combined. It also depends on how we document, assess and judge a work's significance and success, and how it relates to notions such as art, aesthetics, ethics, activism, artistic practice, research, methodologies, the social and the political.

One example of a work with aesthetic integrity that deals with contested public spaces of Gothenburg is the book *Den Urbana Fronten – En dokumentation av makten över staden (The Urban Frontier – A Documentation of the Power over the City)* by photographer Katarina Despotovic and writer/researcher Catharina Thörn (2015). In the format of a photo/artist book, they examine the transformation of Kvillebäcken, a neighbourhood in central Gothenburg. It is a story of how a former industrial area, populated by small businesses, shops and associations was obliterated and converted into a showcase for sustainable urban development. It is also a story about how an everyday place turns into an urban front, a point where a new city should grow at any price. The transformation is part of Gothenburg's long-term change from an industrial port city into a new identity: a strong brand with an exciting and attractive inner-city that attracts investors, businesses and tourists. In a way, the city has cannibalised and eaten the area and its former residents. In the transformation process, image-making was a strategy, creating images of the future as pictures, stories and text.

In the book, the authors tell the story by juxtaposing photos, texts and quotes, leaving space for the reader to formulate the analysis and eventual critique herself. Despotovic and Thörn have spent several years collecting the material and documenting the development. It is an example of working *with* what is there, at the very centre of the events. By presenting photos of the former residents and their activities, side by side with quotes from protocols, marketing campaigns, and statements about the new vision and plan, the material in the book speaks for itself. It becomes obvious that this area holds multiple stories and that the transformation aimed at replacing one class with another. The residents' and the developers' ability to influence the process was uneven when the restructuring took place, but in the book they get equal space. The transformation of Kvillebäcken is a classic example of what Neil Smith describes as a disinvestment, dispossession and gentrification cycle (Smith, 1996).

In the early 2000s, a few years before the transformation and regeneration process started, the city planning department described the area as important.

> Socially, the area is of a fragmented nature … Ethnic associations, gadget markets, bazars, small workshops and motorcycle clubs are intermingled with shops, offices, lunch restaurants and ateliers … Many of the activities available in the area today are run by people who may find it difficult to break into the conventional labour market. The area thus fulfils an important social function and is likely to have a bearing on the growth of new businesses. The activities are therefore of major importance for society.
>
> (Gothenburg City Planning 2002, cited in Despotovic and Thörn, 2015: 79, 173, translated by author)

A few years later, the story changed. In order to legitimize a forced gentrification process, Gothenburg politicians and developers were desperate to erase the history and install a new narrative of the place. It was a process in two steps: first they described the current area in negative terms, and then they presented a solution and a vision of the area's future character (Thörn, 2015: 238–241). Around 2006, the heterogeneous workplaces and the social diversity were no longer seen as resources, but as something untidy and degenerated that needed to be cleaned up. Two parallel stories emerged, one presenting the area as criminal, the other presenting the area as empty. This allowed for the exploitation of the area without taking into account what was already there, neither the buildings nor the people who were active there. A new vision was formulated for a green and sustainable residential area in the middle of the city, targeting an urban and mainly white middle class, which is assumed to be attractive for the city. Today, in 2016, the old buildings and residents are gone, replaced with 2,000 flats, together with shops, cafés and restaurants. In this case, with the municipality's help, a few wealthy players had almost total influence on the planning and subsequent demolition of the area. Those who were active in the area had minimal opportunities to make their voices heard when the municipality was working so closely with the private commercial interests.

The book and the photos unsettle the idea of a single story of the place, and lay bare the economic and political power strategies that are employed in the process. The authors also present an essay that critically examine the development and link it to urban development and gentrification theories. The critique is not delivered from an outside position, but from the very centre of events by just juxtaposing the material itself. Art, in this case photography in combination with text, can provide a material and a way to research and get hold of these ephemeral stories of a place. By publishing a photo/artist/research book with strong aesthetic qualities, they give the material a form and format that can be shared, discussed and create new dialogues.

Returning to the park outside the police station and my work *Guide by Numbers*, the topic of present and absent bodies and voices was a theme that emerged from the material. In the performance, which is in the format of a site-specific guided tour, I am a guide and I have eaten the city's capacity to express *everything* with numbers. I have cannibalised and appropriated both the language and form of the official story of the successful event city, and included hidden, forgotten, and previously un-noticed stories and facts. Through statistics, occupancy rates and *exact* numbers, I describe present and absent bodies in the park when taking visitors on a tour:

> Here are some of the younger trees. They are currently in the juvenile stage, in other words, they are in growth. As you can see, four are alive, one is dead, and so the survival rate here is 80%. They also have fences around them: three out of five function, making the fences' degree of usage 60%.
>
> This compares with the custody centre building over here. There are a total of 185 detention cells, and an average of 151 people are locked up. So the degree of usage, or occupancy rate, is 82%.
>
> Across over here, we have Ullevi Stadium, also known as Ullevi Arena. There, they have a capacity for 75,000 people, and the occupancy rate last year was an average of 30% counting 'bums on seats'. Though, at Ullevi you see the grass pitch very well. That's the whole purpose of the arena's oval design. At a concert, such as Bruce Springsteen, only 4.5% of the tickets were sold with restricted view, and therefore with a lower price. At the opposite side of the park, inside the custody centre, there are 90 detention cells with windows towards the park, in other words, 49% out of the prison's total of windows. But, all windows towards the park have grills so you can't see the grass pitch, which means that 100% of these seats have restricted view.

Those who do not have any windows, nor a view, are those individuals underground. The grass we stand on is an old garrison graveyard. Here lie 12,000–15,000 persons. 950 individuals have been exhumed, making the occupancy rate underground 93%. If you consider that 6,000 persons work in the neighbourhood, and 15,000 are in the ground, you can conclude that 29% of them are above ground, and 71% underground.

(Hedemyr, 2014a)

Using the term occupancy rate, as a way to describe the place, was an appropriation and cannibalisation of the commercial language used to argue for an expansion of arena complexes, office buildings, and hotels in the vicinity of the park. Occupancy rate is mostly used to boast about the commercial success of attracting visitors and tourists. In the process of creating a work for the park, I decided to appropriate the dry language of statistics as a strategy to work *with* the material of the place, and as a way to re-present the statistics in new combinations. In the format of a guided tour in the park, I talked about the absent, dead or locked in, persons that existed right in the middle of this commercial event area. That the park is on top of an old graveyard is not known to the general public. Nor was it known to me when I started to work at the park.

Conclusion: Art digesting public space

As a process of seeing anew and raising questions about the world we live in, I see a potential in works that retain an aesthetic integrity at the same time as they unfold the complexity of a situation and produce new modes of social encounters and critical dialogues. In the complex and unpredictable urban public space, works and strategies have to be tried out, performed and tested at each specific context and site. Artists and artworks that expose and respond to the socially urgent with an aesthetic imagination have the potential to realize the transformative power of understanding and create the urban anew.

In Gothenburg – 'The Event City' – there is always something to eat for cannibals. There are constantly new events that are contesting the borders between reality and the image of that urban reality. The particularities of the political inscribed in this local geography of the event city provides a rich material that can be critically digested and responded to socially, politically and aesthetically. In the business plan 2015–2017 for the destination Go:teborg there is a call:

The world around us is changing rapidly and competition with other cities and regions has increased considerably in recent years. The City of Gothenburg must raise its ambition level and have the courage to invest more proactively to defend and consolidate its position in the visitor industry. This requires visionary efforts that reach beyond the current structure.

(Götburg & Co, 2015: 2)

Cannibals – we now have to raise our creative, critical and aesthetic efforts in order to reach beyond the current structure.

References

Andrade, O. de ([1928] 1991) 'Cannibalist Manifesto'. Translated by L. Bary (1991). *Latin American Literary Review* 19 (38), Pittsburgh, PA: Carnegie-Mellon University, Dept. of Modern Languages) 38–47.

Bishop, C. (2012) 'Participation and spectacle: where are we now?' In N. Thompson (ed.), *Living as Form: Socially Engaged Art from 1991–2011* (First Edition), New York, Cambridge: Creative Time Summit and MIT Press, 34–45.

Cruz, T. (2012) 'Democratizing urbanization and the search for a new civic imagination'. In N. Thompson (ed.), *Living as Form: Socially Engaged Art from 1991–2011* (First Edition), New York, Cambridge: Creative Time Summit and MIT Press, 56–63.

Despotovic, K. and Thörn, C. (2015) *Den Urbana Fronten: En dokumentation av makten över staden*, Göteborg, Sweden: Arkitektur Förlag.

Deutsche, R. (1996) *Evictions: Art and Spatial Politics*, Chicago, IL: Graham Foundation for Advanced Study in the Fine Arts; Cambridge, MA: MIT Press.

Doherty, C. (ed.) (2015) *Public Art (Now): Out of Time Out of Place*, London: Art /Books.

Go:teborg (2015) *The Gothenburg Magic* [Video/trailer]. Available at www.goteborg. com/en/event-organiser/ [Accessed 28 July 2016].

Göteborg & Co. (2015) *Go:teborg Business Plan 2015–2017* [pdf], Gothenburg, Sweden: Göteborg & Co. Available at www.goteborg.com/en/event-organiser/ [Accessed 28 July 2016].

Hedemyr, M. (2014–ongoing). *The Event Series* [Public art project] Gothenburg, Sweden.

Hedemyr, M. (2014a) *Guide by Numbers* [Performance as site specific guided tour] Gothenburg: Motbilder/Counterparts.

Hedemyr, M. (2014b) *Personal Artistic Research Diary* [Notebook], unpublished.

Jönsson, D. (2012) *Estetisk Rensning – Bildstrider i 2000-talets Sverige*, Stockholm, Sweden: 10tal Bok.

Lefebvre, H. ([1974] 1991) *The Production of Space*, Oxford, UK: Blackwell.

Lundström, I. (2012) 'Kalastjuvar'. *Faktum* (119) [online]. Available at http://faktum.se/ kalastjuvar/ [Accessed 28 July 2016].

Malzacher, F. (ed.) (2014) *Truth is concrete – A Handbook for Artistic Strategies in Real Politics*. Berlin: Sternberg Press; Graz: Steirisher Herbst.

Massey, D. (2005). *For Space*, London: SAGE.

Öberg, J. (2013) 'Göte Borg & korruptionen'. *Glänta* (2–3), 22–41.

Sernhede, O. (2016) Göteborg är en stad som glider isär och det blir bara värre. *Göteborgs-Posten* [online] 24 April 2016. Available at: http://www.gp.se/nöje/kultur/krönika-göteborg-är-en-stad-som-glider-isär-och-det-blir-bara-värre-1.199053 [Accessed 21 August 2016].

Smith, N. (1996). *The New Urban Frontier: Gentrification and the Revanchist City*, London: Routledge.

Thörn, C. (2015) 'Urbana frontlinjer'. In Despotovic, K. and Thörn, C. (2015) *Den Urbana Fronten: En dokumentation av makten over staden*, Göteborg: Arkitektur Förlag, 236–247.

12 Our house in the middle of the street

Nela Milic

Introduction

The following chapter is based on my own interaction with the people of Southwark borough in London, particularly in the Elephant and Castle area, and it is drawn from my experience as a participatory artist working with its communities. It proposes that the engagement with 'place' is crucial for art to be transformative because, when art is 'divorced' from place and tries to be exercised elsewhere, this transformation is lost, so the main value of the participatory practice diminishes – the connection to the place. I will be reflecting on the projects I worked on in the last decade, which were situated in the urban environment that started to change in the way many in the Southwark community felt ostracized from, but were severely affected by.

In recognition of previous work I had undertaken, I received an award by the Southwark Arts Forum for the best community art project in 2015. A few months later, I got a job at the University of the Arts' (UAL) London College of Communication in the middle of Elephant and Castle and in the heart of its regeneration. Instead of the famous shopping centre that served as a driver of the local economy for the particularly immigrant community, there will be our new building, surrounded by students' halls, which are already marketed on the previously contested sites.

Where will our students meet the locals, and do their shopping? Would they be able to afford the residencies situated next to the penthouses? How will I teach them about participatory arts, that exposes social and economic disparity, when our institution seems to take part in it by eradicating large parts of spaces for public use? I will try to answer these questions below and crucially, direct the readers towards engaging with these complexities by inviting them to journey through practice with me to understand the plight of a participatory artist and the challenges that arise in the work with the community.

Participatory artwork is destabilizing the juxtaposition of the middle-class art world that is comfortable in most environments and lower-class people who do not traditionally interact with art goers. Bishop calls this experience of discomfort 'antagonism' in relational aesthetics (2004), which takes the sheer presence of the other as an intrusion. She has devised this theory by relying on Rancière's (2006) 'relational aesthetics' and his belief that people think that art has to remove itself from the art world to be valuable in real life and, so, participatory art has to abolish itself as art in order to pursue its goal of social engagement.

In the above mentioned awarded project, this is exemplified by the young people's interaction with the project as they hung around my ad-hoc studio space. When they came to the final show, which consisted of their portraits and other records of the process of the local community engagement, they told me that they had never been to an exhibition before. Surrounded by their neighbours who seemed to have become their extended family rather than by the cultural elite, this introduction to art felt like my greatest achievement in this project. Bishop understands this when she writes that contemporary participatory artists are more interested in creating a socially rewarding experience than establishing a particular aesthetic, so the course of creating a project becomes the product itself (Bishop, 2012).

The links artists make and maintain to further their practice and themselves as inhabitants of the community and incorporate their volatile connections with the city, which are often embodied, consensual, and introspective, are the same as for the projects' participants. The artist, if she is a socially engaged practitioner is compelled to immerse herself in such a challenging environment because being in situations and experiences is the crux of participatory practice. Kester describes this method of engagement as a 'pragmatic openness to site and situation, a willingness to engage with specific cultures and communities in a creative and improvisational manner, a concern with non-hierarchical and participatory art processes, and a critical and self-reflexive relationship to practice itself' (Kester, 2011: 125).

The output of the participatory artist, who works in urban areas, is direct involvement in the dynamic of the city and the actions of its residents. Her or his artistic production is often immediate, revealing and visible, and so, always exposed to critique by the public, but the artist is trained not to be afraid of

Figure 12.1 Young people at the studio (photo: N. Milic).

rejection by the same community she or he wants to work with. This courage, confidence or arrogance, if you will, does not skew the creative expression as the artwork is open to take this in or move in a different direction. Furthermore, artists' parameters can be purely subjective and in that sense, provide case studies for urban and other researchers. Likewise, the artists' work can simply serve the purpose of experiencing the circumstances they put themselves in, playing with the particular art form or method, sharing the skills with the community and vice versa, or just having fun with others.

Most artists are educated for years, so even though the inclusivity and democratic pull of the participatory arts encourages the notion that 'everyone is an artist' (Beuys, 1974), participatory artists know what is needed for the work to resonate with art history, quality of practice and mastership of technique – it was part of their schooling. That does not make the art exclusive, but the art world is and many participatory artists are outside of it because of its traditional rigidity and resistance to collective practices. They tend to be mediators between the elitist art world and inclusive social world, accepting that they will never be embraced by either one. They will still strive to work in both to maintain their practice and challenge the reputable fields, including themselves, the communities they work with and belong to, with a view to better our societies through direct and creative engagement with it.

The 'people' material

In 2012, the Olympics left a desolate land on its site in East London. After initial negotiations about the inclusion of local artists in the production of the artworks alongside the sports amenities, many artists were relocated or evicted from the affordable studios, workshops and community spaces they created since moving into the area years ago. Following the announcement that London would host the Olympics, the rents rose, the housing prices rocketed and a variety of luxury apartments were erected in neighbourhoods around East London. Today's alien landscape – a deserted site that is still under construction is a result of an aesthetic orchestrated by the Olympic committee and city officials who did not want to see 'fridges stuck up on each other in the field' (Sumray, 2015).[1] They preferred clean, simple, corporate visual solutions brought to the borough by the Olympic sponsors like McDonalds. They spoke about the financial revenue achieved through this deal proudly in the Olympic legacy event held at the newly opened Stratford campus of University of East London in 2015.

Four years after the Olympic Games, Hackney council, where the Olympic village was built, has the highest obesity rate among the children in the city,[2] while the artistic heritage of the borough, with the greatest number of artists in Europe,[3] has been replaced by bankers and other high-income generating professionals. The 'native' community was expelled from the area, starting with the Irish traveler population who were previously asked by the council to adjust their mobile life-style by laying bricks to their home.[4] They ditched the trailers and built houses, then they were asked to leave.

Focus E15 mothers still refuse to leave their social housing as they do not want to go elsewhere and consider the flats they have at the moment to be good enough.[5] The new E20 quarter has been bought up by the speculative investors and rich minority who largely live abroad and come to London for occasional weekends that they often spend in Harrods or Westfield shopping centre. They are like gated communities who do not have an interest in participating in their locale, but they paid for their home and they can manage to keep paying by purchasing others ahead of the Londoners who cannot afford to stay on the east side.

Watt, Frediani and Butcher deduce,

> While the rhetoric of regeneration emphasises community consultation, public-private partnerships and the creation of new mixed communities, the reality all too often means displacement, disenfranchisement and marginalization for pre-existing communities that do not benefit from increased investment, or cannot cope with rising property values.
>
> (2016: 13)

Even so, property developers advertise the area as having 'a thriving established community' (Get Living London, 2015), cancelling responsibility for taking part in diminishing precisely that community they describe. Wainwright in *The Guardian* in August 2016 reveals: 'all the neighbourhoods planned so far can only muster an average of 30% affordable housing', leaving the ones with the capital to dictate the establishment and the type of community spirit.

We, the artists who moved out of the area as the cost for parking spots for the summer of the Olympics rose to £800 a month, grieved the loss of our homes and work places in a different way, mostly by trying not to be on the site during the Olympics. We had nothing to celebrate. So, I spent my time at the artists' residency south of the river Thames where I encountered a gentler rate of regeneration, slowly picking up pace and surfacing the issues I was used to experiencing in the east.

The day I arrived at the old Victorian row of houses in Southwark, I was approached by a couple of barefoot women in long dresses who were watering the flowers in their garden. Those 'fairies' that opened their door for me had recently moved here from Hackney. We, the dispersed, found each other. What will I be doing during my artist's residency, I asked. They tell me that I am to find the connection between the local architecture and the memory of this private space that used to belong to a family. They gave me a tour of the area in silence and I saw an abandoned council estate, bustling food market, a few Georgian and Victorian houses and the vast Burgess Park. So, I was to link the past and the present in my participatory artwork, the form of which I was yet to find. I looked at the traces the previous artists had left in the space and I decided to let the community choose what to do here.

In the meantime, I researched the locality and recorded the process of re-development, especially its signposts – posters, adverts and marketing banners hanging from the building sites. I noticed that people had disappeared from the images of Southwark councils', developers' and even commissioned art projects on regeneration. While I was surrounded by people on the street, who seem to be part of the busy public transport as the traffic lights, buses and bus stops were everywhere, the developers came up with the statement: 'The Elephant & Castle has no soul ... there is no community here'[6] (Deck, 2012). Lend Lease company, which is a development partner to the council, promotes its transformations of the area in its brochures as a drive to bring Elephant and Castle back to 'its rightful zone 1'[7] aesthetics (Lend Lease, 2012). Considering that it could have meant Kings Cross, Westminster, Oxford Street or Soho as central city areas, I chose to be driven in the artwork by the residents' desires about their neighbourhood to see if they matched those of the grand renovators.

South via east

I engaged with the constituency that I saw on the streets by opening the ground floor window of the two-bedroom flat that became the art studio I was working in. This junction of public and private space over the course of the residency served as a vessel between me at home and the people on the street. The outside world was looking at the indoor one and I quickly realised that even though contained and enclosed, the home offers many unexpected occurrences and linkages. For example, the public did not expect anyone to say 'hello' through the window, let alone invite the viewing of a private space from the street.

Likewise, I did not expect anyone to stop in front of the window and talk to me, let alone about their private lives. The boundary between socially divided spaces was collapsed and the public walked straight through that parting as if it never existed. By releasing the contours of the space separated by different notions of living (flat or house, ground or last floor, small-build or high rise) a critical artscape was formed, and I could pursue adaption of it over time, tracing and accelerating various forms of community engagement through the emerging collaborative praxis. With this relation, I was feeling the connection that the community had with their built environment.

We laughed, cried and shared goods – drinks, CDs, dogs even. I organised a film screening and did a different engagement practice every day, starting with running a photo booth, accepting local history tours I was taken on by the people, opening a beauty parlour on a day when the local beauticians went to Notting Hill carnival, setting up a newsagent for reading papers and sharing the news, and inaugurating a window gallery showing the photos I took because that day it was raining. I was going to be a mending service before I finally opened the doors of this project space and let people in the way we normally enter someone's home – through the door rather than have an exchange through the window. For ten days of my residence, 'our house was truly in the middle of the street'[8] and everyone came to it.

First, I have been intrigued by the idea that this section of quite ruined houses is run by Westminster housing cooperative.[9] People were allowed to live there if they took care of the property that was in transition – the lease holders had an

Figure 12.2 Busy street (photo: E. Vikstrom).

interest in giving it another function or knocking it down in four to five years. One of the reasons for this was to avoid squatters moving in – whose lifestyle cannot be controlled, or boarding the property to avoid drug addicts living there, about whom the local community was complaining to the landlord.

Then the housing cooperative brokered a contract with the creatives, which stipulated that the current residents are only 'guardians' of the site, but they were forward looking enough to commit to an extra flat that they used as artists' studio, which I was invited to utilize for the residency. They called it 'Syzygy' – a synthesis of energies. The curatorial programme they initiated was concerned with re-appropriation of their home, which I used for exploration of urban development. I focused on the links between the past successes in the positioning of material environments for community life and the current removal of those physical reminders of our heritage as well as erasing the whole communities with it.

I had worked in the area before the residency, through artistic engagement with Cooltan Arts.[10] My task was to produce a wall artwork with the elderly who spent time in the local day centre for vulnerable older people. I held two hour sessions once a week, exchanging stories about the past lives of the residents of Stones End[11] centre who delivered them in chronological order that resembled a family album. Even though they were geographically positioned differently – some came to the United Kingdom from Jamaica, the others from Eastern Europe or the north – they all narrated their lives by years. I clustered them into decades I assigned to tree branches when we decided that the artwork would be a family tree of all the people at the centre.

This three meters-long installation consisted of a drawing as well as the photographs the residents contributed to provide the content for the visual hierarchy the tree naturally displayed – the early years in the lives of the residents were at the beginning of the last century, settling at small branches developing from the trunk of the tree, and the big global events like the wars showed when the tree blew its full crown. The pictures were replaceable, so the residents could navigate their own stories through the tree if they wished as the time went by.

The encounter with the elderly, who generously shared their wisdom, the support of the centre's staff and the context of the site was inspiring. I kept in touch with them as well as with Cooltan Arts following its evolution from the barracks in a street passage to the new office space on the Walworth Road in Elephant and Castle, which suits a leading local arts charity in the field of mental health. The plans for the new building were on the table, but the realization of that construction has failed. The board dismissed the chief executive who set up the charity decades ago and the funds amassed for the development of the new space were spent. Many members left the organisation in allegiance with the chief and now have nowhere to go. Even though the workshops, classes and training continue via regular forms of Cooltan Arts programme – walks, talks and educative sessions – the charity is not as welcoming as it used to be. Its last gathering at Drapers Hall revealed a wounded community, divided between the management and the need for creative engagement.

While the infighting was publicised with every members' account, I could not escape thinking how all of this started with the revelation of the plans for the new build. Since I spent time at Syzygy space, I have learnt a lot about the struggles in the neighbourhood with the process of regeneration. It did not happen at a fast pace as it did in east London due to the Olympic schedule and it was not as concentrated on one site, but spread around, targeting first social housing. The well-known story of decay of the famous Heygate Estate was unravelling while I was in residency, but there was so much written about it that I stayed clear of the site that had started generating income through art projects interested in urban destruction.

Furthermore, the council was banking on artistic tourism, not only from the fees for various film productions (even Hollywood found its way into the estate[12]), but it transported second-hand shipping containers from Scandinavia to the site with the vision that the artists would wish to use them as their studios whilst they explored the desolation of the location. The windowless structures appeared where there were previously builder's workshops and many did not dream they were to be for rent, let alone at £1000 a month, a sum local artists could not afford to pay even for their own homes in the area. Even the artists who worked hard to set up their own spaces, like the choreographer Siobhan Davis, managed to erect her new building in close vicinity to the Elephant and Castle station just as the battle for the Heygate was in its height, making the higher-profiled artists look like the scavengers of the estate's carcass.

The trouble was that the passers-by who became project participants at the studio as they chose to involve themselves in it further, started talking about Aylesbury Estate down the road in the same way as the case of the Heygate had been articulated. I initially thought that media, policy makers and community representatives provide the language of the issues surrounding the politics of the estate redevelopment so that the people will adopt it when they are communicating what is happening elsewhere, but then I saw the notices, council reports and builders' releases and realized that it is all the same – there is nothing to be told in a new way. People are witnessing the same damage to their immediate environment, which is clearly working for the regeneration agents since almost every London community involved in the struggle for their habitat loses the fight after a number of meetings, protests and actions and the costly apartments appear on the sites of their once affordable homes. No wonder then that Lend Lease, an Australian property giant, is also redeveloping E20 after obtaining a catastrophic reputation with the Heygate site.

In *Estate, a Reverie*[13] (2015), a film by Andrea Luka Zimmerman, the attempt of creative management by the Hackney council was exposed when they employed a bright orange cover for the boarded flats they planned to demolish and re-build for the residents who were gradually moved from one place to another across the canal. For Andrea, who lived in one of the flats when this venture began, this childish colouring of what was a traumatic event of leaving one's home was a pitiful interpretation of the arts projects, which socially engaged artists were producing locally. The pull of participatory art and its democratic aura opened itself up so much that the authorities thought they could do it too, but failed when Andrea and her collaborators in the *I am Here*[14] (2009) project replaced the boards with the portraits of the residents, defying the representation of the estate by its council as an abandoned, derelict, worthless construction and instead illustrated that people lived there and were proud of that.

Collaboration

As I was already crushed by the Olympic landscape changes in the east, I focused on the work with the community in 'the Elephant', insuring they appreciated and enjoyed the process of introduction to, management and temporary ownership of the space I was residing in. I have invested time in listening to their stories of the relationship with the locality and I responded to it visually and in a performative way. I decorated my window every morning with a cloth and tea cups and positioned a bench outside it, so the people understood that this is a space for replenishment and contemplation over a 'cuppa'. I put on an apron and a notice on the building, 'free tea and biscuits', and started shouting from the window at the people across the street who were accustomed to avoiding my side of the road where the trouble used to be.

Interested in what was happening and drawn to the environment that now looked welcoming as a domesticated space, most of the passers-by came to meet me. I explained that I was new in the area and I wanted to know the local people. My camera was disclosing that I was recording this process, however, I was not making a big deal about my profession, but more about the space I was working in. Confused somewhat with the fact that my space was empty, the people started suggesting what I could do in there. The advice ranged from painting the walls to making it homely to highlighting the history of the house by establishing a museum with the objects found in the space and in the locality. Even though a display of objects would be a classic choice for a visual artist and I already had a good

Figure 12.3 Window conversations (photo: L. Sayarer).

start – a part of the old Victorian pram excavated from under the stairs – I opted for a more participatory approach, providing services that I had heard from the community were in demand.

Responding to those requests gifted our space with a variety of local people who had previously kept themselves separate according to their habitat – business owners did not mix with the local residents unless there was an economic exchange, people from houses were reluctant to make friends with the inhabitants of the estates, and young people did not hang around where the elderly were. This is not an unusual living practice in London, but the spaces for the community coming together were disappearing with the new developments looking to profit from every square metre, and were mostly not including community spaces in their new buildings. With the vanishing of old estates, community spaces were going away and local businesses, which struggled to keep themselves afloat, had less and less time to spend on the people who were not asking for their service and just want to be somewhere collectively.

So, I arranged for my studio to be a communal place, providing daily space for the residents' use. Within it, I offered some activities that I believed gave joy to the visitors and they were mostly encouraging spending time together. B[15] came every day from the start of the project, eager to try out anything suggested and to tell stories of her own creative endeavours. S was also around since she discovered the space, admitting that she struggles to discipline her now regular companion – a dog that her doctor advised her to get after she tried to end her life. M across the road first just eyed out the commotion around the studio as it was a reminder of the space's history, but soon she came down to ask what it was all about and we had her from then on regular tea breaks.

G arrived one day with the disability service users who had a music class in their centre at the end of the street, and they provided a concert outside the window as he thought that I was house-bound considering that I did most of my work indoors. I was also given stuff people thought that I could use and I obtained a new hairdresser. Even the postman enquired about the local addresses in my studio. Occasionally, I was simply ignored – people sat at the bench in front of my window to finish their phone conversations or eat a sandwich on their lunch break.

Residents from the Aylesbury estate were increasing in numbers as the word spread that 'the artist' in the area was providing 'things' without charge. One discovered that I would take their picture and that they could pick it up on the opening of the exhibition, so they went home, dressed up, and invited friends and family members to a group portrait session. The photo booth attracted everybody, even the least likely locals, like a finance officer of the theatrical supplier across the road who kept away from my camera for a few days. Delighted with her portrait at the exhibition, which I put up on the last day of residency, she stayed around until the end of the party. I stuck the portraits on the wall on the 'private view' and asked people to peel theirs off if they wanted to take them away. The exhibition went up during the day and it came down that night.

In all exchanges, I was creating a relationship with the space and with the community that also generated inter-subjective links, I was trying to gain an understanding of how my co-producers live locally and what happens when the communal space is missing. I was not driven by a particular outcome, potentially inviting in utopia – although to know that active involvement in the framing of the residents' surroundings after our collaborative project would be appreciated – the participation itself was the goal. I have reflected on what has been conveyed at the time and set up the activities in the studio as a response to it, but I was not the only one supporting the residents with their necessities. 56a coop[16] that was a bit further away ran a bike workshop, a food store with produce grown by the local people, and it was also the home of a world-renowned anarchist bookshop.

One of the representatives of the art world on the site was visibly excited with the exhibition and openly stated how much he enjoyed the evening of 'coming together'. As I communicated this to the residency curators, who were later invited to contribute to the education programme in the mainstream local gallery, I also expressed my disappointment with the fact that the curator, who was clearly inspired by the work and had come to see it a few times before, had not thought of me exhibiting it as part of his official gallery line-up. Maybe he already knew that such work would not be selected for it by the whole organisation's team.

Sentenced to the education stream of the art world, most participatory artists do not have the opportunity to develop this important practice as they often have to pull out of it in order to make a living. Furthermore, the need to create the work often stops artists from seeing the circumstances within which it has to be done, infringing their professional moral standards that should be the core of socially engaged practice because participatory artists are traditionally more interested in producing an ethical connection with participants than nurturing critical judgment (Cohen, 2012). For me, that in itself is a critical judgment of the society the artists engage with.

The artist's space

For Bishop, strategies of 'art in the public interest', intervention or over-identification could be ruled out as precisely 'unethical' (Bishop, 2012). Kester is also concerned with how artists can facilitate change by creating projects, appropriated by governments for 'soft social engineering', to slowly dismantle the welfare state through inspiring citizens to take responsibility for their own community development in deprived areas as a compensation for economic redistribution (Bishop, 2012).

However, participatory artists tend to work in the areas where the community is already formed in order to support low-income families by swapping goods, taking care of each other's children and elderly, and generally doing things as a group, rather than individuals. As the community has already nurtured a sense of the shared good, the participatory artist can organically direct the collective artwork that makes the practice perfectly suitable for those urban environments, so much so that oversights are made in terms of socio-political conditions within which the art is created. Even though it is the responsibility of the artist to tune-in to the surroundings rather than ignore them because she might be taken by the practice which encompasses that knowledge; artists, as part of the community also rely on the community for support in relation to the understanding of those environments.

Community too has the power to accept or refuse to be part of both – creative or political engagement. For example, as I saw in my residency, the community does not show an interest in living in the penthouses parachuted into their neighbourhood, but is rather drawn to the ground structures that offer a space for free-reign. Zone one, as imagined by the developers, is not for community taking and they have expressed that in the many surveys and focus groups conducted as part of the local regeneration process, which they recalled throughout our project.

Furthermore, theoretical discourse on the use of art or artists in the city neglects the account of practitioners – their reflections on the production, personal and professional politics, the interaction with commissioning bodies and their sense of impact of work on the audience. Most investigators of the ever altering 'urban' do not feel the need to transform a static spectator to an active participant (Van Heeswijk, 2001), while participatory artists insist on it and often pose the question of how their own work can be run as a cooperative process.

Artists take responsibility for making community agendas noticeable and producing the work encapsulating the artistic briefs, but when there is position (institutions, corporation, market) and opposition (community), dividing public space into a battleground between those powers surrounding it, the artist is pushed to make a choice in working for or against the community. Most artists are nowadays caught in the middle of these two sides, hence, making work without a political stance becomes increasingly difficult even for the most skilled negotiators – artists and community members who often lose the arguments or even credibility in the hands of more powerful players that have stronger financial, media and regulatory capacities.

In a talk at 2015's Art and Journalism festival (hosted by London College of Communication), the artist Jeremy Deller explained his process of creating *The Battle of Orgreaves*[17] (2001) about the miners' strike in 1994, which ended up being a well-known participatory art piece in the United Kingdom and around the world. Imagined initially as a performance inspired by documentary television footage the artist saw as a young man, Deller choreographed a re-enactment and filmed it.

The theatrical ability to capture the event and consequently provide its legacy for further viewing, research and reference, allowed the artwork to travel from its local setting, where it was powerful because of the authenticity of re-enactment performed on site by the miners who were then on strike, to the international sphere where the focus shifted to its aesthetics, rather than political moment in British history. When the work entered the global art market, the forces of the market itself managed to override the political dimension of the piece that came back to the United Kingdom less charged with the local radical impetus. The performative form of the work, the number of the people in the field, and the act of battle moved the audiences more than the decline of the working classes in Thatcher's Britain the event captured.

A further problem that the global art flow creates is in the artwork's concept, which can become more important if interpreted by the curators than the artists' rationale of the work. This challenge arises because artistic management often accompanies urban development plans initiating profitable creative cities, cultural enterprises and mega productions. Such economic wealth reduces the artist herself to a servant with a useful skill to communicate with people. Equally, the artist can choose to manipulate communities, to be complicit and support the governing structures she claims to challenge and strives to change.

However informal the outcome of their production seems, the artists who are working in a participatory field are talking about upholding similar professional standards that range from small practices in their neighbourhood to intercontinental, costly, grand-scale projects. Even though the experts' rhetoric around the ethics in both categories of artwork – small or large – sounds the same, a practice taken away from the local community misses large parts of its meaning when its 'original' context is dislocated. It can become diluted into the connective, dialogic, interventionist realm of socially engaged art that only appears more collaborative and democratic because it has more partnerships, funds and participants, impressing with the scale rather than the quality of work with people. It tends to be formalized and rigid, prescribed by its agenda as much as official, canonised, mainstream art is in its approach to the significance of disciplines within art itself.

This is not to assert that participatory quality comes only with the small, contained, 'next door' projects, but covering the vast territory with the artwork or aiming at drawing in the whole area around it evokes interest in addressing social or political needs, which is not necessarily a job for the artist. Indeed, art can satisfy a desire for community engagement thoroughly and express what cannot be done as successfully by other means, but governing organisations have a duty to plan and deliver social programmes beside art provision whilst participatory artists have trouble being accepted as formal art practitioners in their own sector.

This is because social immersiveness often directs the mode of expression and its aesthetics, seemingly less controlled, less independent and less artist-led than other author-led practices where the artist is the sole author of the artwork she creates. An additional difficulty for the artist is the fact that the art displaced from the site where the artist mainly works becomes valued by its relationship with other works in the new place.

To find critical validation in art outside art history is already strenuous, let alone by engrossing in the practice of participatory arts explored here in relation to the contribution it makes to the complexities of urban development. Its spatial context disperses creative energy for the purpose of mobilizing community rather than focusing on an aesthetic product, even though for Rancière, the aesthetic is in itself political (2006). Still, 'Energy yes, quality no!' (Hirschhorn, 2007) is just one paradigm that evaluators of 'conversational art' (Bhabha, 1998) contemplate – the other is assessing how collaborations are performed and judged. In the examples used in this chapter, I surveyed both of those categories – the value of community engagement and a display of that engagement as well as the projects' artistic products (performances, exhibitions, activities) created in Elephant and Castle. I argued that public participation is figured through the discourse of space that can be understood only with a view of time when the public engages with it to build a sense of belonging to that space and so, make a 'place'. Taking away the locality within which a place is created, participatory arts loses its strongest dimension – its community making.

Conclusion

Within geography, artscapes are interpreted as places for disruption although they might reconcile often abandoned communities. In the course of regeneration, those close-knit groups of people are usually, through gentrification and redevelopment, moved away from each other, eradicating the once established spirit of being together. In such positionality, participatory artworks upset new orders set up by the state, corporative and monetary forces that influence creation and erection of urban environments because they are often re-establishing communes and encouraging citizens to have agency in negotiations about their habitat.

Through equipping those communities with tools for resilience by urging them to act together against ambivalent perspectives posed by urban developers and city administration that support profitable rather than civic, historically appropriate and socially relevant projects, artists influence the notion of public sphere from its material to philosophical constellations. However, they can also take part in strengthening the authority that capital and political power brings to occupations and utilisations of public space.

The compromised stance of the public artist in cases against public space comes from the artist's acceptance to justify the interference of divergent actors that have an interest in appropriation of that space. By providing the artwork on the queried site, artists are mediating tension between the openness of space for public use and private, business or statutory authorities; encircling that space that can close that space upon becoming its stakeholders. Their community art has 'an often paternalistic or judgmental relationship to local cultures, forms of knowledge, and social patterns ... with little regard for the conditions pertaining at a given site' (Kester 2011: 125). In this way, participatory artists and frequently government-funded initiatives in the queried spaces have the same goal – to foster social cohesion.

Temporary built environments, transitory places and mobile architecture embrace diverse publics – community groups, visitors, creatives and experiences enabling a cycle of short-term, experimental, festival-like interactions within a long-term programme. This overarching method ensures the advent of the new praxis of co-production as it is long enough to devise artworks and record their social and spatial reception (Cumberlidge and Musgrave, 2007). Known as 'artivism', 'creative resistance' or 'artistic resilience', the artistic practice immersed in geographical discourses addresses the problems with disciplinary distinctions, balances the variations in access to the space founded by different agents and complements community engagement and social movements. This complex artistic creation with people is often aimed at challenging or shaping anew and with unforeseen knowledge gained throughout the projects, a particular landscape they inhabit or use for work, leisure, meetings and so on.

Artists and residents who are active in the moulding of their environment, organize and participate in workshops, discussions and planning procedures, with a view to modelling their upcoming living space. These city makers create the topography of the site by intervening in public spaces as they are constructed, conducting an ethnography and adjusting to the changes in the area under construction, taking into account the new residents and future generations (Kwon, 2002). The result is that the research depicts transformation of the city and its communities and so, contributes to the comprehension of the critical artscape.

Temporal dimension is also very important in participatory art. The significance of accounting for 'places through their history' is such that O'Neill (2010) calls upon the idea (inspired by Latour's 'Cohabitation Time', 2005) of 'public time' initiated by the duration facet of art. O'Neill understands democratic space as the time spent together publicly with some common objectives. He is interpreting time as the principle where durational modes of public engagement contribute to the production of that space by welcoming social forms of art-making.

The constitution of time makes the space – a place where co-productions emerge. Those social encounters with artists denote participation to an unfolding, lived experience that is therefore, creating temporary cohabitations. These sequences of time engineer creativity and the unknown and unexpected outcomes that are often unquantifiable. The public manifestations generated this way are an accumulation of gathered collective contribution.

Temporality drives mobility, transitioning states within which people are and encouraging 'becoming' (Bergson, 1946; Deleuze, 1995) by evolving their actions over time. As the relationships are always in flux, nothing will occur again in the same way, but this pregnancy with a difference is a testament to the change that participatory artists look out for, devising various types of participation according to the time. Participants too, seek to modify themselves with time during the course of the project to gain personal value with their participation. This worth is embedded in curatorial form as participation experience that requires the inclusion of abilities and wishes the involved community might have of the work.

In this time-guided way, participatory projects conceive the 'place' – a negotiable, hybrid, dynamic, evasive and open construct inviting the social bearings that contribute to the creation of its coherent identity, however problematic collective identity can be. The relational artworks fit in it perfectly, exposing research-based results that can only correspond with their specific context, public (audience and participants) and location which is passing, instable, contested and with plural trajectories. Consequently, revealing inequalities and class differences with these projects makes a will for social change difficult to ignore.

In circumstances that affect the repurposing of public space and its function, it is important to hold on to the memory of space, converting the desire for domination into an alternative geographic imagination (Gregory, 1995). By identifying themselves as the propertied class that dictates social contention, the landowners remember the past in their own way, typifying it and preserving it over the others (Said, 2000). This 'imaginative geography' is the tentative construction of space less invested in territory and its inhabitants and more involved in charting memories, maps, narratives and physical edifices.

Such space is a constellation of different paths of activity, which is how Massey saw 'relational place-making' in her book *For Space* (2005). For her, spatial and social are related, so the space is always in production. Therefore, space has a potential for equality in the urban development, but we must not forget that it can be politicized and reliant on the capital and hence, its history (Massey, 2005).

'Place-making' is also less concerned with the traditional venues of artistic production like museums and galleries and finds its showing platforms in community centres, residential quarters, the street and the Internet. In this manner, critical artscape is part of a larger domain of public realm. Critical artscape expands that realm in order to import subversive strategies within it. This tactic is three-fold: to use the opportunity to present the issues that are not covered by mainstream artists and cultural institutions; to obtain finances for the regular sustenance of practice and practitioners; and to elevate the work of artists whose work is too controversial to be supported.

For the success of this approach, it is necessary that we arrest landscape with memory that can only live with people in it because they are able to interpret and represent it as its 'rightful' inheritors. Participatory arts projects allow for this to be done with the sense of ownership not only of that space, but its story, and so it is identity that symbolizes the people of a 'place'.

Notes

1 Richard Sumray, formerly chief executive of London International Sport and consultant to LOCOG provided this comment at the event *Debating the London 2012 Olympic Legacy: Implications for Future Olympic Cities* at UEL in 2015.
2 http://digital.nhs.uk/article/5240/Highest-rates-of-childhood-obesity-in-those-living-in-deprived-areas.

3 http://hackneywick.org.
4 http://issueswithoutborders.com/archives/39.
5 https://focuse15.org.
6 Rob Deck, former Lend Lease Project Director of The Elephant regeneration 'The Murder of the Elephant', *Southwark Notes – Whose Regeneration?* Available at southwarknotes. wordpress.com/2016/06/13/the-murder-of-the-elephant/ [Accessed: 4.8.2016].
7 'One The Elephant is designed to symbolise the restoration of The Elephant to its rightful place as a thriving zone 1 neighbourhood' – Lend Lease brochure *One the Elephant.* Available at www.onetheelephant.com/apartments/~/media/Developments/UK/STM/ Documents/EC%201the%20Elephantemail%20friendly.pdf [Accessed 5.8.2016].
8 The song 'Our house' was made famous by the UK rock band Madness that released an album *The Rise and Fall* in 1982. The catchy refrain from that song contains the lyrics: 'Our house in the middle of the street'.
9 www.westminsterhousingcoop.org.
10 www.cooltanarts.org.uk.
11 www.ageuk.org.uk/lewishamandsouthwark/our-services/stones-end-day-centre/.
12 www.24dash.com/news/housing/2011-09-28-brad-pitt-films-zombie-movie-on-hey-gate-estate.
13 www.fugitiveimages.org.uk/projects/estatefilm/.
14 www.iamhere.org.uk.
15 I have kept the anonymity of the members of the community I worked with by assigning only their initial to their name as I did not seek the clearance for publishing of the reflection on the work from them at the time of the project. Since many have moved away since the project was realized, I do not have an opportunity to do so retrospectively.
16 http://www.56a.org.uk.
17 Commissioned by Artangel through the open call in 1998.

References

Bergson, H. (1946) *The Creative Mind: An Introduction to Metaphysics.* Trans. Andison M. L., New York: Philosophical Library, 129–132.

Beuys, J. (1974) Statement, 1st published in English in Tisdall, C. Art into Society, Society into Art London: ICA, 48.

Bhabha, H. (1998) 'Conversational Art'. In Jacobs, M. J. and Brenson, M. (eds.) *Conversations at the Castle: Changing Audiences and Contemporary Art*, Cambridge: MIT Press: 38–47.

Bishop, C. (2006) 'The Social Turn: Collaborations and its Discontents'. *Artforum*, February: 178–183.

Bishop, C. (2006) 'Introduction // Viewers as Producers'. In Bishop, C. (ed.) *Participation*, Cambridge, MA and London: MIT Press and Whitechapel.

Bishop, C. (2012) *Artificial Hells: Participatory Art and the Politics of Spectatorship*, London and New York: Verso.

Bishop, C. (2004) 'Antagonism and Relational Aesthetics'. *October.* Fall No. 110: 51–79.

Cohen, R. (1958) 'David Hume's Experimental Method and the Theory of Taste'. ELH, XXV, no. 4, 276-277.

Cresswell, T. (2004) *Place: A Short Introduction*, London: Blackwell Publishing.

Cumberlidge, C. and Musgrave, L. (2007) 'Introduction'. In Clare Cumberlidge and Lucy Musgrave (eds.) *Design and Landscape for People*, London: Thames and Hudson, 15.

Deck, R. 'The Murder of the Elephant', Southwark Notes – Whose Regeneration? Available at southwarknotes. https://southwarknotes.wordpress.com/2016/06/13/the-murder-of-the-elephant/ [Accessed: 4.8.2016].

Deleuze, G. (1995) 'Control and Becoming'. In *Negotiations*. New York: Columbia University Press.

Deleuze, G. (2004) *Desert Islands of Other Texts* 1953–1974. Trans. Sylvére Lotringer. Los Angeles, CA: Semiotexte.

Frieling, R. (2008) 'Towards Participation in Art'. In Rudolf Frieling (ed.) *The Art of Participation*. London and San Francisco: San Francisco Museum of Modern Art and Thames & Hudson.

Get Living London. (2015) Available at www.getlivinglondon.com/gll-neighbourhood.aspx [Accessed: 24.8.2016].

Gregory, D. (1995) 'Imaginative geographies'. *Progress in Human Geography* 19 (4): 447–485.

Guerlac, S. (2006) *Thinking in Time: An Introduction to Henri Bergson*, New York: Cornell University.

Hirschhorn, T. (2007) *Where do I stand? What do I want? Quaderni a spirale e carta Spiral notebooks and paper*. Paris, France: Collection de Bruin-Heijn, Galerie Chantal Crousel.

Kester, G. H. (2011) *The One and the Many: Contemporary Collaborative Art in a Global Context*, Durham and London: Duke University Press, 178.

Kwon, M. (2002) *One Place After Another: Site-Specific Art and Locational Identity*. Cambridge, MA: MIT Press, 52.

Latour, B. (2005) 'Bruno Latour: From Realpolitik to Dingpolitik – or how to make things public'. *Pavilion Journal for Politics and Culture*. Available at pavilionmagazine.org/bruno-latour-from-realpolitik-to-dingpolitik-or-how-to-make-things-public/ [Accessed: 25.8.2016].

Massey, D. (2005) *For Space*. London/Thousand Oaklands/New Delhi: Sage.

O'Neill, P. (2010) *Art and Urban Development*. Original in English *De rive* 39.

Ranciere, J. (2006) 'Problems and Transformations in Critical Art'. In Claire Bishop (ed.) *Participation*, Cambridge MA/London: MIT Press and Whitechapel, 90.

Said, E. (2000) 'Invention, Memory, and Place'. *Critical Inquiry* 26 (2) Winter, The University of Chicago Press, 175–192.

Sharky, C. (2008) *Here comes everybody: the power of organizing without organizing*. Penguin Press.

Van Heeswijk, J. (2001) 'Fleeting Images of Community'. In Annette W. Balkema and Henk Slager (eds.), *Exploding Aesthetics*, Lier en Boog, Series of Philosophy of Art and Art Theory 16. Amsterdam/Atlanta: Rodopia, 178.

Watt, P. and Frediani, A. and Butcher, M. (2016) 'Regeneration Realities' *Urban Pamphleteer* 2. Available at www.ucl.ac.uk/urbanlab/research/urban-pamphleteer/UrbanPamphleteer_2.pdf [Accessed: 15/8/2016].

Wainwright, O. (2016) 'London's Olympic legacy: a suburb on steroids, a cacophony of luxury stumps'. *The Guardian*, 3 August 2016.

13 Crowd creations

Interpreting occupy art in Hong Kong's Umbrella Movement

Sampson Wong

The genre of 'occupy art'?

As an urban researcher and an artist who actively participated in Hong Kong's Umbrella Movement in 2014,[1] I have been frequently asked to reflect on and think through the phenomenon of art flourishing overwhelmingly in the three occupied zones.[2] The intensity of art's presence in this particular occupy movement has been widely acknowledged.[3] It is also argued that among the itinerant occupy movements across the globe, the Umbrella Movement established the strongest connection among art, politics and urban space (Law 2016; Lee 2015a; Pang 2016). How should one understand this connection? How can one adequately conceptualize and theorize the strong presence of art in Hong Kong's Umbrella Movement? These are the questions this chapter attempts to address.

The Umbrella Movement Visual Archive, co-founded by myself, preserved a collection of more than 300 art and creative objects and over 1,000 posters generated in the occupied zones.[4] As one of the initiators of the project, I have often been asked what kind of objects were kept in the archive. On the one hand, I had to clarify that the most iconic and well-known sculpture, barricades and murals were not kept in the archive, on the other hand, I struggled to name the objects by their established genres, such as sculpture, poster, drawings and so on. I often answered that they are the more mundane and ordinary creations of the people who participated in the occupy movement. In part, a substantial part of the art produced cannot be adequately named by the usual categories of art objects, while collective efforts and continuous creations in the occupied zone as a whole can be conceptualized as a specific phenomenon and a genre of political art yet to be understood.

In this chapter, I provide an initial framework to analyse the 'occupy art' in the Umbrella Movement by focusing on what I call 'crowd creations' that emerged in the three occupied zones. I then suggest that by focusing on the crowd creations we are able to grasp the particularity of occupy art in Hong Kong. In turn, this chapter also suggests that only by focusing on the particularity of occupy art in Hong Kong's Umbrella Movement, are we able to understand the protest that took place in 2014.

The spectacular settings of the Umbrella Movement, which involved long stretches of roads and highways in core urban areas of the city, together with various pieces of iconic art and interventions, dominantly captured the attention of global media and audiences.[5] Apart from these images and visual elements that defined the events, hidden from view were the spontaneous, more ordinary, mundane, small-scale and continuously emerging aesthetic participation of the citizens, who stayed in the occupy zones daily and fused their daily lives with the movement for months. Contrary to the overall settings and iconic interventions, the latter type of aesthetic interventions is less demanding and easily adopted by the general public who participated in the occupation. This chapter highlights the latter and provisionally names them as 'crowd creations'. Artists and cultural practitioners were active in many occupy movements across the world, thus producing what can generally be conceptualized under the banner of 'occupy art' (McKee 2016; Mitchell 2012). However, greatly differentiating the occupy art in various occupy contexts and Hong Kong's Umbrella Movement was the intensity and strong presence of crowd creations in the latter.[6] The current chapter suggests that the strong presence of occupy art in the Umbrella Movement that seemingly let it stand out as a particularly 'creative' occupy movement, should be supplemented by the specifically strong presence and uniquely vast amount of crowd creations. On the one hand, such an understanding facilitates our search for the actual meanings and implications of 'a strong presence' of occupy art in the Umbrella Movement, on the other hand such analysis points to the importance of revisiting the connections of occupy art to the specific cultural and political dynamics within each occupy movement. In short, under the same 'genre' of occupy art, and their common nature and function across various occupy movements, there could be specific constellation of art production within each occupy movement that reveals more about its unique spatial and political conditions. The current chapter takes up Hong Kong's Umbrella Movement as an example to carry out such an inquiry; it seeks to enrich the literature on occupy art by framing art in occupy as a lens to look at the distinctiveness of an occupy movement.[7]

I take up the inquiry as outlined by first formulating a general conceptual distinction between the professional and centralized form of collective artistic interventions in social movement, and spontaneous and decentralized crowd creations. Second, I further articulate the distinction between the two forms of aesthetic interventions by situating them in a framework concerning what I conceptualized as exogenous and endogenous functions of occupy art. Focusing on crowd creations in the Umbrella Movement invites us to look at the internal dynamics of the protest; it leads us to look at how occupy art was deeply entangled with the cultural and political specificities of Hong Kong's occupy movement.

Centralized projects (occupy art 1) versus decentralized crowd creations (occupy art 2)

If one were allowed to enter the occupied zones of the Umbrella Movement just after the participants were evicted, one would have been astonished by the material contents that remained. A major amount of traces and objects constituting the materiality, or objects produced by participants who have had no contact with the professional art sector and had not made art before they spent time in the occupied zones. The productions and practices involved in yielding the objects were not part of any collective plan initiated by political leaders or artists; they were the aggregation of highly spontaneous and self-motivated individual effort. The multitude of spontaneous and self-motivated aesthetic effort, involving citizens influencing each other in-situ, have altogether composed the texture and atmosphere of the occupied zones.

Contrary to the collective action creatively planned by political leaders, visual interventions consciously produced by key figures and celebrity in popular culture industry, and artworks made by professional artists, crowd creations are more impromptu, spontaneous and reflective of how the occupied spatial environment was exerting its influences to the creators. While images of these creations were often readily circulating in mass media and dominated people's perception of the Umbrella Movement, it should also be noted that their nature could be highly participatory and they could engage a large group of participants. Taken as a whole, what differentiated them from crowd creations was that they had a highly structured and pre-defined pattern of creator-spectator or creator-participant relationship, mainly designed by professional artists and political leaders.[8] Contrary, crowd creations describe small-scale and atomized self-motivated projects undertaken by individuals; most of them were participants of the Umbrella Movement who fused their daily lives and occupy lives gradually as the occupy zones remained for almost three months (Chan 2015). While it was common for these spontaneous creations interacted to generate more creations (Lee 2014), these were bottom-up crossing paths and encounters rather than constructed situations. Compared to art and interventions created under the expectation of attracting spectators and involving a calculated participatory mechanism, crowd creations in the occupied zone served better in capturing the mass's desire to voice and its ephemeral will to create and act spontaneously in occupy times.

By now I have attempted to make a conceptual distinction between two modes of aesthetic interventions, coexisting in the occupied zones, that were often simply referred to as 'occupy art'. If we look closely, the art and creative projects that constitute our strong visual impression about the image of the protest were mainly highly structured interventions that maintained a centralized control of how spectator and participant react to the interventions. They were mostly projects initiated by artists and political leaders. For example, the celebrated *Lennon Wall* project, initiated by local artists, was constructed by a call for all participants to write and draw on standardized post-it paper[9] (see Figure 13.1). In this chapter I name similar initiatives that tightly prefigured and structured recruited participants and audience's mode of participation, as 'centralized' art (I also refer to them as 'occupy art 1' to clarify my argument).

The more spontaneous and individualized aesthetic practice, collectively forming what I have called 'crowded creations', were characterized by their decentralized mode of creation (I also refer to them as 'occupy art 2'). Occupy art 2 was sometimes made in response to prominent examples of occupy art 1. For instance, after the *Lennon Wall* was created and was established as the artistic icon, there were numerous 'mini-*Lennon Walls*' created by individuals using the surfaces of

Figure 13.1 Protest encampment in central Hong Kong.

card boxes as walls to stick post-it messages on them. These imitated versions of mini card boxes were in varying sizes and designed differently, since each of them was a self-motivated, individual project. In stark contrast to the centralized aesthetic of the iconic *Lennon Wall*, the many small card boxes with post-its on them were collectively manifesting a kind of decentralized aesthetics as we can observe the spontaneity and variations among them. In the coming parts of the chapter, I shall discuss more examples of occupy art 2.

One may suggest that it is not difficult to accept that the two types of artistic interventions coexist in occupy movements across the world, and one may question what we gain by thinking through occupy art with such distinction. This chapter argues that in the specific context of Hong Kong's Umbrella Movement, the conceptual distinction of occupy art 1 and occupy art 2, with an emphasis on the latter, allows us to understand the following particular dynamics of the movement (See Table 13.1):

a   Since from early on the artistic aspect of the Umbrella Movement was high-lighted by the global and local media, paradoxically, the overwhelming attention placed on art highlighted occupy art 1 in an unprecedented way. As such, the tendency of highlighting the strong presence of art in the Umbrella Movement reinforced a mainstream understanding of occupy art solely as occupy art 1. Occupy art 2 was marginalized. Although one may argue that the tendency to overlook occupy art 2 is common across all occupy move-ments, the paradox was that a strong media focus on art as such would lead to a further overlooking of occupy art 2.

Table 13.1 Conceptual distinction between centralized projects and crowd creations

	Occupy art	
	Occupy art 1	*Occupy art 2*
Conceptual distinction	Centralized mode of creation and aesthetics	Decentralized mode of creation and aesthetics
Producer	Professional artist and political leader	Majority of participants of the occupy movement
Image circulation in various forms of media	Abundant	Scant
Mode of interaction	Structured and calculated	Spontaneous and heterogeneous
Naming	**Centralized Projects**	**Crowd Creations**

b   However, if one overlooked occupy art 2, one would be overlooking a core
    aspect of the Umbrella Movement: its radical decentralization and the lack
    of dialogical mechanism. While many occupy movements celebrated leader-
    less protest and direct democracy, it is important to note that the Umbrella
    Movement – with its three separated occupied zones in different urban core
    areas – was radically decentralized, but the attempt to form a decentralized
    and deliberative mechanism was absent.[10] Throughout the two and a half
    months, experiments and attempts to form new kinds of leadership or new
    kinds of political dialogue were unsuccessful. Interestingly, in that situation
    participants began to devote themselves to crowd creations. In short, the vast
    amount of creative objects produced and the great number of participants
    choosing to engage in the Umbrella Movement, spontaneously by participat-
    ing in crowd creation, were reflective of the specific outcome of a decentral-
    ized protest that emerged in Hong Kong. Although the current chapter cannot
    extend its focus to the general political culture and culture of protest partici-
    pation in Hong Kong, the culture of participation in the Umbrella Movement
    can also be situated within the passive and polite politics (Ho 2000) in protest
    and a fundamental lack of precedent practices related to deliberation.

c   As a decentralized protest in which people hope to maintain a leaderless
    occupy movement, the consequential mode of participation and the choice
    of cultural practices of the Umbrella Movement's participants were also
    reflective of the specific trajectory and the 'life' of the occupy movement.
    Spatially, the Umbrella Movement comprised three separated occupy zones
    while they were all happening in major traffic routes and highways. To sus-
    tain an occupy movement in this spatial condition requires sustained efforts
    of repair and maintenance, and crowd creations often combine repair and
    maintenance with art and creativity. Temporally, as the occupy prolonged to
    up to two and half months and a sense of frustration emerged, the participants
    continued to search for actions to gain a sense of continuation and engage-
    ment. Rather than simply looking to be involved in centralized projects,
    many were spontaneously engaging in crowd creations. They were created

to construct meanings for a major portion of the mundane and waiting time they went through daily since they had no particular political task to take up in most of the days. As such, if one has to grasp what 'participating in the umbrella movement' means, an unavoidable route is to focus on occupy art 2.

Based on my role as an active participant, observer and archivist in the Umbrella Movement, this chapter contributes to the discussion on art and politics in the context of occupy art by highlighting crowd creations as an overlooked component. As such, it explains the specifically central role played by crowd creations in the Umbrella Movement and preliminarily outlined what they revealed about the movement. I further do so by distinguishing centralized projects and crowd creations by what I call their 'exogenous' and 'endogenous' functions respectively: it is argued that looking at the endogenous functions of occupy art (expressed through occupy art 2) allowed one to think through the particularity of the protest, compared to other occupy movements that emerged globally.

Centralized artistic projects and exogenous functions

This discussion allows us to grasp the particularity and potential meaning of 'art in the Umbrella Movement', if we acknowledge that aesthetic activities went way beyond the 'standard package' of aesthetic participation contributing to the construction of occupied zones as spectacles. If the closer connection between art and politics in Hong Kong means that more citizens have been willing to participate in collective creative interventions requiring their minimal participation, and more of these interventions were initiated by active artists and politicians, establishing them as core elements in social movements, then for most citizens engaging with art and politics in protest means either anticipating, following and spreading images of these interventions (taking up the role as an 'attentive spectator'), or being part of a top-down and pre-designed collective creative action for it to be completed (taking up the role as a relatively passive partaker).

The continuation of this development was witnessed in the occupied zones, where pre-designed and well planned artistic interventions, inviting occupiers to take part, were proliferating; throughout the Umbrella Movement sustained anticipation and attention was drawn to the artistic sphere. However, crowd creations in the occupied zones reflected a development that cannot simply be subsumed under the pre-Umbrella Movement pattern stated above. The participants, be they 'inhabitants' of the occupied zones or not, devoted efforts in setting their own creative and artistic agenda: rather than taking part in other orchestrated interventions, they actively initiated micro and often mundane creative actions and the production of small aesthetic objects.

Many of them will find the idea of shaping artistic interventions as spectacles irrelevant and had not been conscious of making images for circulation in the mass media. The mode of citizen aesthetic participation in politics has transformed into one that prioritizes agency, active participation and the act of creation itself, over image-making and its effect. In the occupied zones, one can witness a wave of artistic production that involved creation for creation's sake, citizens turned from attentive spectator (Lau and Kuan 1995; Lee and Chan 2008) and minimal participant in the overlapping terrain of art and politics, to active creators that are more preoccupied by the will to express and participate in a more autonomous and bottom-up fashion. Participation in politics through art has a new and unfamiliar meaning for the Hong Kong citizens, and it allowed crowd creations to fill the occupied zones.

If there has been a novel phenomenon of much more active and wide-spread artistic creation emerging in the Umbrella Movement, such novelty and its implication for political participation in Hong Kong has not been well captured by conventional interpretations. This chapter argues that the inadequate interpretations, offered by most discourses on art in the Umbrella Movement, privilege an 'externalist' reading of the creative activities and artistic objects, in part influenced by how occupy art has been generally approached globally. An externalist framework interprets artistic practices mainly as means for the occupiers to communicate and interact with the world external to the occupied zones. Three typical ways to characterized protest art and occupy-art are outlined; their popularity in part explains why the novelty of political art (in the Umbrella Movement) to Hong Kong was not acknowledged and why the significance of crowd creations was overlooked.

First, influenced by recent theorization of protest art and disobedient objects (Flood and Grindon 2014), creativity in occupy movements is often highlighted as efforts contributing to defending the occupied zones and enhancing the strength of individuals and the temporary infrastructures. However, in Hong Kong's case the fusing of creativity and art with the defensive structures is exceptional among most creative initiatives. Even though the use of bamboo scaffolding in construction of barricades was once heavily reported by the media (Bradsher 2014), similar creations on the ground was scant. While in many cities creativity has been applied to design of personal equipment for defending police violence, social movement in Hong Kong as a whole is largely nonviolent and a more militant wing of activist is largely absent. The type of creativity manifested in defensive initiative seen in other occupy movements is in a minority. Also, it is indeed when the occupied zones were more stable and defensive calculation was minimized, artistic activities blossomed.

Second, protest art is often conceived as political propaganda and an apparatus of convincement to mobilize citizens who are politically neutral or hesitant by an alternative way. However, one only has to take a quick look at the majority of art being produced in the occupied zones in Hong Kong to find out that they seldom contain messages that speak to the general public or the occupiers' political opponents. The crowd creations we find in the Umbrella Movement are mostly not fulfilling the propagandist function, instead they reflect the desire of the occupiers to speak to other fellow participants of the movement.

Third, as the most spectacular and well known artworks and interventions created by professionally trained artists and political leaders were characterized as landmarks of each occupied zone, the media attention they received gave people the illusion that most artistic products were aesthetically successful enough to help shape the public image of the Umbrella Movement and reinforce the pride and aesthetic superiority of the pro-democracy camp. For the decentralized, crowd-creations produced spontaneously, not only were they never included in maps of the Umbrella Movement surveying 'key landmarks' and media reportage, they were also not generating images that are utilitarian to broad portrayals of the occupy zones from an 'aerial' perspective.

Decentralized crowd creations and endogenous functions

The exogenous framework dominating our perception of art in the occupied zones is supported by the public's general understanding of protest art's defensive, persuasive and imagery functions. The abundance of crowd creations (as were most of the art produced in the occupied zones) prompt us to analyse them from an endogenous perspective: the specificity of the almost negligible protest art is that they tend to be simple daily means for the occupiers to gain a sense of participation and empowerment in the prolonged and decentralized occupy movement.[11] Specific to Hong Kong's context, the majority of art and popular participation witnessed in the Umbrella Movement also signals a qualitatively different form of artistic-political participation, as more citizens experimented with the practice of being creators instead of passive participants or spectators.

Shifting one's attention away from the 'landmarks' and a number of the most revered art pieces, and the minority of art that fitted an exogenous understanding, allows researchers to rediscover the neglected crowd creations filling up the occupied zones. An endogenous framework also implies that creative practices embodied intrinsic values for the citizens-turned-creators, the act of participating in the Umbrella Movement through art was fundamentally meaningful to their temporary life world. For the many who had participated in art-making and micro-creative projects, art can be a communication tool within small and intimate groups; it can be about spending and wasting time, as most occupiers were bored most of the time. Art was thus a means for occupy participants to re-order their disrupted normal daily lives, and also a means to endow meanings to their new and atypical lives and to make sense of them. Art allowed many ordinary citizens to literally leave their marks in the occupied zones, and to turn the small spaces they occupied or frequently stayed from generic spaces to meaningful and decorated places. When boredom became a sense of powerlessness, art was also a means of empowerment, as participants created to remind themselves and others that 'at least something can be done'.

The vast amount of crowd creations produced and actions initiated by the participants, often anonymously made and unnoticed, nevertheless defined the ethos of collective joy, co-creation and popular participation throughout the occupying times. It may not be easy to pin down the categories and genres of crowd creations that appeared in the occupied zones, yet it suffices to list some types of creative routines that played out most of the days.

While banners are core to politics and visuality in the everyday urbanscape and all social movement performances, it should be pointed out that the number of hand written and hand drawn banners is still remarkable in Hong Kong's history.

During the nights, numerous banners were put up and re-hung in the occupied zones; during late nights one could witness small groups of occupiers creating new banners and choosing preferred locations to mount them. Contrary to factory produced banners, banners existing in the occupied zones exhibited craftsmanship and were usually products of slow collective efforts. Some of the banners might not even be seen for a few days, and were covered by a second layer of banners. The abundance of banners also reflects how the varied and layered urban landscapes are influencing each occupied zone. Researchers may not have been able to count and record the number and content of banners produced throughout the Umbrella Movement, but it should be acknowledged that they altogether consti tuted a core aspect of the occupied landscape, and production and maintenance of banners were activities central to many occupiers' daily lives.

On almost all kinds of vertical surfaces within the occupied zones, posters were covering every inch of the space (Figures 13.2 and 13.3). Perhaps it is also impossible to roughly estimate the number of unrepeated designs that had appeared. In terms of creation and production, the threshold of knowledge and resources for making posters was relatively low. Yet, it is still shocking for one to notice the vast number of participants in the Umbrella Movement who insisted on designing their own posters and printing them at home or office; the low budget productions enabled the emergence of large-scale democracy walls, which should be considered one of the most outstanding visual elements of the occupy movement. Apart from posters printed with paper and home printers, paper boxes and markers were the main materials used for writing messages in the occupied zones. Placards made using hand-torn cardboard were placed in all corners; they formed the most primitive form of physical message board system, even though the ones with more elaborated design and function were more eye-catching.

Figure 13.2 Banners were constantly hung on bridges within the main occupied zone.

Figure 13.3 Apart from walls of posters, many participants of the Umbrella Movement engaged in creative interventions in situ daily.

Some of these placards were decoration and signage attached to tents in the occupied zones. The creative efforts devoted to modification and beautification of tents were equally effective, when compared to the almost unlimited production of banners and posters. The efforts to add aesthetic quality to tents and people's habitat were extended to all kinds of occupied spaces and objects within the spaces. Public spaces, barricades, signage, bus stops and so on, had been transformed and hijacked. It was difficult for people who stepped into one of the occupied zones to re-imagine how the spaces were in pre-occupy times. Functions of spaces were transformed, and the will to enhance the livability of the occupied zones triggered immense creativity. One can simply take a closer look at how lawn was transformed into farm, how public toilets became truly public as the public took care of maintenance and decoration, how street libraries emerged and how new routes were created, to sense the creativity circulating in the zones. Importantly, many of the transformations were not well planned and designed; the phenomenon can rather be seen as results of accumulated crowd creative efforts without orchestration. To put it in Anja Kanngieser's (2013) way, the concerted creative efforts were political experiments, taking the making of worlds as their aims.

Last but not least, all kinds of crafts and micro art projects were put up; in some of the days occupied zones were a marketplace for taking and making different objects. They existed in the form of stalls and booths with varied size. Throughout the Umbrella Movement, on each night, there were up to 100 active booths in which people gathered to form micro communities with usually a dozen participants communicating intimately. Participants of the Umbrella Movements learned origami and folded uncountable umbrellas made of paper, one could find stalls for learning and distributing these origami umbrellas, while workshops creating other installation art pieces by assembling the umbrellas together would take place alongside the stalls. The continuous appropriation of key symbols and messages, the emergence of stalls for craft-making and objects distribution and the establishment of workshops experimenting innovative ways to make use of the crafted products, formed a specific ecology of crowd creations. This specific ecology was utterly different from the dominant form of protest art in pre-Umbrella Movement times; the mode of participation for the mass was altered. The intensity of bottom-up creations illustrated an important aspect related to art and politics in the Umbrella Movement, namely that the number of creators of objects and initiators of micro-art projects would not be much less than the number of spectators of political art and participants in collective artistic interventions. For a large portion of occupiers and participants of the Umbrella Movement, part of the experience of participation involved first-hand creations and artistic efforts. The majority of art produced in the Umbrella Movement was not the most reported and noticed pieces, but the vast pool of crowd creations that were almost taken for granted.

Conclusion

This chapter attempts to make the case for the significance of crowd creations in the Umbrella Movement. It argues that if there is an urgency to look at art in the Umbrella Movement, it should be about the phenomenon of popular participation and creations in the occupied zones. In short, this essay answered the question of 'what to look at' by pointing to the subtle and daily act of decentralized creative practices, undertaken by citizens participating in the Umbrella Movement. It was these crowd creations creating the ethos, the overlooked joyful undertone, and the fluid sense of empowerment in the occupied zones. This essay has only initially explored the possibly important political implications, if we focus on the specific aspect of art and politics that manifested in the Umbrella Movement. Some of the preliminary arguments, yet to be further substantiated and articulated, are summarized below as a conclusion to the essay.

First, the essay has pointed out that citizens' participation in the overlapping terrain of art and politics, commonly took the form of being spectators or participants in collective actions requiring minimal participation. More citizens were transformed into active and autonomous creative agents in the Umbrella Movement; it is yet to be seen whether the paradigm shift in the mode of aesthetic participation will reshape the role of art in social movements and protests in post-Umbrella Movement interventions.

Second, if we adjust our focus and pay attention to the majority of art produced in the occupied zones, some of the key dimensions of the Umbrella Movement can be reassessed. The meaning of participation may not only be altered in the context of art and politics, but also in the overall context of the occupy movement. Without a detailed analysis of the aesthetic participations, research on citizens' participation in the movement would not be complete.

Finally, the transformation in mode of participation within the context of creativity in protests, may also allow us to speculate on how the Hong Kong people's mode of political participation can be re-conceptualized as a whole. After all, what are the political implications if citizens had realized their potential creative agencies and were willing to be creators and setting aesthetic agendas? These are the preliminary issues for addressing the question of 'why look at art', after addressing the theoretically complex question of 'what to look at'.

Notes

1 Apart from producing the artwork 'Add Oil Machine' and co-founding the Umbrella Movement visual archive, I also curated two exhibitions about art in the Umbrella Movement in New Zealand and in Hong Kong.
2 Three urban cores of Hong Kong were occupied, namely Admiralty, Mongkok and Causeway Bay. Among all three of them, areas being occupied were mainly major traffic roads.
3 There is even an entry 'art of the Umbrella Movement' on Wikipedia detailing the creative aspect of the protest event.
4 To know more about the archive, please see Qin (2016) and access www.facebook.com/umbrellaarchive/.
5 For examples of media reportage, please see Paul (2014); Sheehan (2014); Young (2015).
6 The naming of 'crowd creation' and my focus on it, is in part drawing on Murphy and O'Driscoll's (2015) conceptualization of 'visual ephemera'.
7 I have also outlined a general and wider research agenda concerning art and the Umbrella Movement in Wong (2016). This chapter articulates what I believed to be the fundamental conceptual framework in understanding art in the Umbrella Movement.
8 For a discussion on the many imaginations concerning mechanism of participation and politicizing art, see Bishop (2012).
9 See Dapiran (2015)'s discussion.
10 However, one researcher has also argued that there were communication activities that served the function of civic education, see Lee (2015b).
11 For a discussion on their digital counterpart, see Lee and Chan (2016).

References

Bishop, C. (2012) *Artificial Hells: Participatory Art and the Politics of Spectatorship*, New York: Verso.
Bradsher, K. (13 October 2014) 'Building a Bamboo Bulwark Against the Hong Kong Police'. *The New York Times*. Retrieved from www.nytimes.com.
Chan, W. L. (2015) *Activists in the Umbrella Movement: A Phenomenology Study of Students in the University of Hong Kong*, HKU Scholars Hub: Undergraduate thesis. Retrieved from hub.hku.hk/handle/10722/219939.
Dapiran, A. (2015, March and April) 'The dilemma of illegal art'. *Art Asia Pacific*, 92. Retrieved from www.artasiapacific.com.
Flood C. and Grindon G. (2014) *Disobedient Objects*, V&A Publishing, London.
Ho, K. L. (2000) *Polite Politics: A Sociological Analysis of an Urban Protest in Hong Kong*, Burlington, VT: Ashgate.
Kanngieser, A. (2013) *Experimental politics and the making of worlds*, Burlington, VT: Ashgate.

Lau, S. K. and Kuan, H. C. (1995) 'The attentive spectators: political participation of the Hong Kong Chinese'. *Journal of Northeast Asian Studies*, 14 (3): 3–24.

Law, S. M. S. (2016) *Concerning Umbrella: Creation, Affect, Memory*. Online research publication. Retrieved from www.commons.ln.edu.hk/vs_faculty_work/5/.

Lee, E. (2015a) 'Space of disobedience: a visual document of the Umbrella Movement in Hong Kong'. *Inter-Asia Cultural Studies*, 16 (3): 367–379.

Lee, L. F. (2015b) 'Social movement as civic education: communication activities and understanding of civil disobedience in the Umbrella Movement'. *Chinese Journal of Communication*, 8 (4): 393–411.

Lee, W. K. (2014) 'Xi Jinping at the "Occupy" Sites. Derivative Works and Participatory Propaganda from Hong Kong's Umbrella Movement', *Trans Asia Photographic Review*, 6 (1).

Lee, L. F. and Chan, J. M. (2008) 'Making sense of participation: The political culture of pro-democracy demonstrators in Hong Kong'. *The China Quarterly*, 193: 84–101.

Lee, L. F. and Chan, J. M. (2016) 'Digital media activities and mode of participation in a protest campaign: a study of the Umbrella Movement'. In *Information, Communication & Society*, 19 (1): 4–22.

McKee, Y. (2016) *Strike Art: Contemporary Art and the Post-Occupy Condition*. New York: Verso.

Mitchell, W. J. T. (2012) 'Image, space, revolution: The arts of occupation'. *Critical Inquiry*, 39 (1): 8–32.

Murphy, K. D. and O'Driscoll, S. (2015) 'The Art/History of Resistance Visual Ephemera in Public Space'. *Space and Culture*, 18 (4): 328–351.

Pang, L. (2016) 'Arendt in Hong Kong: Occupy, Participatory Art, and Place-Making'. *Cultural Politics*, 12 (2): 155–172.

Paul, K. (1 October 2014) 'The Powerful Protest Art Behind Hong Kong's "Umbrella Revolution", *Mashable*. Retrieved from www.mashable.com.

Qin, A. (13 May 2016) 'Keeping Hong Kong Protest Art Alive Means Not Mothballing It'. *The New York Times*. Retrieved from www.nytimes.com.

Sheehan, M. (21 October 2014) 'The art of the Umbrella Movement'. *The Huffington Post*. Retrieved from www.huffingtonpost.com.

Wong, S. (2016) 'On the Umbrella Movement and Visual Arts: from the occupied areas to the art field and the social world'. *Hong Kong Visual Arts Yearbook 2015*, 218–247.

Young, D. (24 September 2015) 'Of a particular time and place': In praise of the Umbrella Movement's iconic artwork. *The Hong Kong Free Press*. Retrieved from www.hong-kongfp.com.

14 #Beirut, urbanism and me

Framing the city through space, identity and conflict

Luisa Bravo

Introduction

This chapter is a journey through the vibrant and fascinating art-based expressions in the city of Beirut, laying as a meaningful presence on the urban environment, strongly permeating its everyday life. Beirut, among the oldest continuously inhabited human settlements in the world, is a city that never dies: throughout its long history, dating back to Phoenician times, the city survived many natural catastrophes, like earthquakes, tidal waves, floods, fire accidents, and a fifteen-year-long civil war (1975–1990) that devastated the city centre. As the Lebanese poet Nadia Tuéni wrote (1979), 'Beirut has died a thousand of times, and been reborn a thousand of times'. Today the central district, the ancient heart of the Roman Berytus (Jones Hall, 2004), has been mostly rebuilt, while many building sites are transforming surrounding neighbourhoods. But different forces from the civil society are generating a new, significant turmoil: at the end of August 2015, a citizen-based movement called 'You Stink' motivated thousands of Lebanese citizens, protesting after a severe garbage crisis, to occupy the streets and squares in the downtown district and the Parliament's offices. This action of civil disobedience came as a consequence of a long period of political instability and of increased tension after the influx of over one million Syrian refugees, since 2011, in a country of barely four million people, that disastrously impacted on the Lebanese society and economy, significantly affecting the housing policy, the labour market and the everyday urbanism. In this social and political disarray, Lebanese artists started to populate the city with beautiful works, such as murals, graffiti and performances, actively taking part in the process of reconstruction of the city identity, strongly reclaiming public spaces from private interests.

This chapter is the result of facts that I witnessed, of encounters and interviews with local residents, especially youngsters, journalists, politicians, scholars and artists. While the civil war destroyed much of the physical environment and any kind of evidence from the past, people are vehicles of memories, stories, events, passing from one generation to another. This human navigation is able to unveil a multitude of emotional 'Beyroutes' (Studio Beirut, 2016), an imaginative, subjective, bottom-up urban culture, overlapping with the official city.

In my journey of understanding, Beirut is neither completely Eastern nor Western: these two faces coexist and constantly contradict each other, while a proud attempt to establish a pure Lebanese identity is boldly raising up. Reacting to a political weight, highly impacting on society, culture and urbanity, open-minded citizens are revamping national pride and reclaiming their right to the city, intertwining their voices with local activists and non-governmental organisations (NGOs), working in collaboration with artists, who are persistently making art as a form of resistance.

#Beirut, urbanism and me

I visited Lebanon for the first time in 1997; I was involved as an architecture student in an international program, the Prince of Wales' Urban Design Task Force, supported by the Lebanese Prime Minister Rafik Hariri. At that time, Beirut was still recovering from the urban and social devastation of the civil war, but a very ambitious plan to reconstruct and revitalize the city centre was already taking shape, promoted by 'Solidere' (an acronym of *Société Libanaise pour le Développement et la Reconstruction de Beyrouth*), a private company set in place by the government in 1994. This plan was somehow controversial, for being much more interested in the tangible qualities of architecture and urban design, through a top-down/market-led approach, and less to memories and identities of different communities, generating a gap between intention and realisation (Hanson, 1997). The driving concept of the Task Force[1] was to tackle the post-war issue of 'reconnecting the city', using a community participation approach and process-oriented urban design strategy, referring not just to the downtown district but also to surrounding neighbourhoods, in a kind of comprehensive urban preservation policy, taking into account also the peculiar character of the city as a whole. Although being involved in a very stimulating environment, a unique and intense educational experience, I couldn't feel the cultural-based extraordinary power of the city, probably due to my lack of experience and to the fact that I was applying my European education to a non-Western context (Saliba, 1997), that was to me largely unknown.

As a grown-up scholar, I visited Lebanon again in 2014 and later, in 2015, I was appointed as Visiting Assistant Professor at the Lebanese American University (LAU) in Beirut. My work, as an academic researcher, was primarily aimed at studying the city's post-war architecture and urban transformation, with a focus on public space. Being an Italian and taking advantage of my previous experience, I was confident at first to find similarities related to the larger common culture of the Mediterranean basin and thus I decided to apply a comparative urbanism approach (Robinson, 2011; Robinson, 2016). But even with a wide access to a vast literature (in Arabic, English and French, the official languages of the country) available on site, I soon realized that most of the knowledge in Lebanon is deeply 'context-dependant', somehow distant from established theoretical knowledge (Flyvbjerg, 2006); it is a path through rumours and hearsays, local books, newspapers and unofficial documents and English written blogs run by locals, observations and epiphanies. My effort to map Beirut stumbled into a layered infrastructure, made of fragmented and contradictory patterns, in a tangled context of politics, power, conflict and poverty.

My perspective as a woman had also some kind of specific, different taste. Despite the cosmopolitan and the freewheeling female behaviour that differentiates Lebanon from other Arab countries, I had the feeling I was being perceived by the locals as a weird Italian lady, for not being married, for not having kids, for driving a car in what many call a crazy and unsafe traffic jungle, and for living alone in a furnished one-bedroom apartment in Hamra, the predominantly Muslim neighbourhood of the city: with my independent attitude, I was breaking many of the unwritten codes of conduct that Lebanese society requests of women (Seidman, 2011). Even if the Lebanese government is promoting equal rights for women, especially in the education sector, for the staff – academic instructors and administrative officers – and for students (but with relevant tuition fees, not easily accessible – Lebanon has just one public university and thirty-one private universities and institutions[2]), it doesn't have a civil code regulating personal status law for women. A report published in 2015 by Human Rights Watch 'reveals a clear pattern of women from all sects being treated worse than men when it comes to accessing divorce and primary care for their children'.[3] Some organizations, like the League for Lebanese Women's Rights, are raising awareness on gender issues, especially with regard to enabling women to reach decision-making political positions.

Living in Lebanon for eight months gave me the opportunity to directly experience everyday life, to pause, to observe and listen, to explore and navigate its prosperity, to confront myself with misunderstandings and misinterpretations, preconceptions and hardship. From a Western perspective, especially for an Italian used to walking, living in Beirut hurts, because the Arab everyday urbanism hurts: with the chaotic disorder of the traffic, with cars moving everywhere, in every possible direction, with the impossibility of walking, due to the lack of a continuous sidewalk or a designed pedestrian path, with the dust coming from countless construction sites, reshaping empty lots and abandoned buildings, with the intolerable smell of garbage, with taxi drivers beeping the horn constantly, with the dependence on a decent, but expensive, Internet connection. It hurts because it affects all activities at every hour, of the day and the night, and this uncomfortable feeling remains stuck to your skin and can make you mad. But everyday urbanism is rich at the same time, with the shades and the colourful lights in the sky at sunset, with the dense humanity along the streets, sitting on every kind of chair, ubiquitous in almost every existing sidewalk of the city, with the smiles and kisses of friends and foreigners that you encounter on the way, with the food and the celebration of conviviality in thousands of restaurants and cafes where people meet and enjoy their time, smoking arguileh and drinking arak. Urbanism in Beirut is made of a bad and good taste mixed together, in a way that it is difficult to describe: it is intangible but powerful, highly fascinating but fragile. I started to understand the complexity of the Arab culture and its hidden meanings only when I detached myself from the schemes and perspectives of my Western understanding, opening to a sphere of multiplicity, even contradictory, voices, desires, emotions and dreams. I found the perfect definition of Beirut's mood in a recently published book: 'Nothing is constant in Beirut. Certainly not dreams.

But despair isn't constant either. Beirut is a city to be loved and hated a thousand times a day. Every day. It is exhausting, but it is also beautiful. You will never feel more extinct or more alive than you will in a day in Beirut. And it'll probably stay that way' (Mashallah News and AMI, 2014: 10). This book, produced by an Arab collective and independent online platform, Mashallah News, is an anthology of less-told or forgotten extraordinary stories of ordinary people, made of nostalgia and beauty, imagination and identity, desperation and fate.

Having the opportunity to work with Lebanese and Arab students at LAU, I asked them to share with me opinions, beliefs, experiences on Beirut and Lebanon, to discuss and deliver their own image of the city, beyond the visible dimension, exploring its potential made of cultural meanings, literature, memories, lifestyles, fashion, trends, food, tourism, social practices of appropriation and transformation; to put into the equation perception and senses in the experience of the urban taste affected by cars, cabs, shops, scents, sounds, views, chairs, light and shadow.

In my design studios, I had to train students in architecture and urban research and to guide them in the design of new public buildings and public spaces in Lebanon. My task was to give students tools to imagine how Beirut and the country could have been. I didn't want to superimpose my Western approach, I was much more interested in understanding and embracing Arab urbanism. I gave my senior students an assignment to write a short essay, 'My city, urbanism and me': I asked them to explain their personal approach to the city, how they see it, how they feel it, how they dream it. That was an extraordinary opportunity for me to see Beirut and Lebanon through their eyes: in their words, I could understand that invisible wounded world, made of pride, desire, hate and love and of an endless stream of emotions.

While in Beirut, I had the chance to attend the theatrical performance '*Hkeelee*' (*Talk to Me*) by Leila Buck, a Lebanese American actress, writer and teaching artist. Exploring citizenship, family life, immigration and loss, from the period of the civil war to a new life in America, she described what Lebanese people suffered, what they hold onto, what they let go, and how those choices came to shape who they are. In Lebanon, as in other divided territories, violence, fear and a profound sense of loss affected residents long after the end of the conflict (Khalaf, 1993; Sawalha, 2010). Listening to Leila's voice, my understanding of Lebanese multiple, fascinating, intangible cultural layers expanded enormously. This ephemeral yet pervasive dimension is the only Beirut that really exists, despite decades of urban devastation.

Beirut, the global city of the Middle East

In less than twenty years, Solidere completely rebuilt the Beirut Central District (BCD), an area of about 200 hectares, through a massive urban reconstruction program estimated to have cost tens of billions of dollars, involving star architects and leading international firms (Bravo, 2017). Resembling the Western model of historic districts, the BCD was conceived as a well-designed system of public spaces, pedestrian paths and preserved heritage, with luxury shops at the ground floor. The urban renewal was based mainly on mixed-use rather than zoning, with a specific intention to create a diffused public realm, giving emphasis to an aestheticizing vision, with the main aim to bring to life once again the character of the pre-war city and the special nature of the place, 'its history, association with sea and mountains, economic role in the region, and links with east and west' (Gavin and Maluf, 1996: 13).

The Beirut Souks (see Figure 14.1) sitting on what was the place of the old souk, opened to the public in 2009, and is today an extraordinarily attractive shopping mall, the most sold image of the city and a must-see place, but something that most of Beirutis can't afford (the average monthly income of the population, according to the World Bank's country profile of Lebanon[4] is about $1,000). Following the Solidere's motto 'ancient city of the future', Beirut was voted the number one destination to visit by *The New York Times* in 2009.[5]

A quite ferocious criticism is largely common by Beirutis, both from those generations who lived during the conflict and the new ones, about the reconstruction process after the war. Solidere was accused of using harassment and intimidation to drive the original residents out of the BCD – former residents and business owners were compensated with shares, rather than cash, at what many claimed was well below the true value.[6] The company used a public/private model: taking ownership of the whole centre in one single plot – with the exclusion of places of worship, governmental headquarters and a few other buildings – it could only act on governmental approval of the master plan and the commitment to build infrastructures and public spaces, getting in exchange a twenty-five-year concession – then extended in 2005 to seventy-five-years – on the rights to exploit the buildings, to manage the services and the permit to build and manage the new waterfront. Local

Figure 14.1 Beirut Souks, designed by Rafael Moneo and Kavin Dash (opened in 2009), serving today as the main retail attraction of the Solidere district. Picture by the author.

owners had the option to keep their property and return the shares, but with the obligation to restore their buildings following Solidere's strict preservation codes. Almost 50 per cent of the buyers of new developments are expatriate Lebanese, but also Arab citizens and companies from the Gulf area, so most of the Beirutis perceive the BCD as an urban precinct designed for those living abroad. The huge real estate business promoted by Solidere generated several counter effects: in less than six years, the price of new constructions reached peaks of 400 per cent, so that the downtown district became an elitist enclave, strangely deserted on a normal daily basis, because of the absence of permanent residents. The gentrification process then affected surrounding districts: according to a study produced in 2014 by RAMCO, a Lebanese real estate advisory agency, the cost of one square meter in Beirut is equivalent to ten times the minimum wage, while an average price for a 250sqm apartment in Beirut is nowadays over $1 million (USD). Beirutis can no longer afford to live in their own city.

Habib Battah, a Lebanese investigative journalist, filmmaker and media critic, in a 2014 issue of *Al Jazeera Digital Magazine*, published a detailed report entitled 'Erasing memory in downtown Beirut. Has the redevelopment of the city's former hub eroded its history and privatised its future?'[7] In the article, he clearly explains how BCD, which was intended to be the opportunity for social classes to mix, became a kind of no man's land for rich people, as a result of political and economic affairs at the basis of the establishment of Solidere. Battah is editor of *Beirut Report*, a blog documenting the other side of official information, giving evidence to what is hidden or not told, paying particular attention to activists' role in influencing decision-making in Lebanon and the Middle East. He became quite popular after he gave a talk at TEDx Beirut, at the end of 2014, entitled 'Welcome to Lebanon',[8] which is the most common thing you hear people say in Beirut when something is broken, like the Internet, sewage system or electricity, that is seemingly impossible to fix. In his talk, Battah addressed the lack of transparency and accountability from politicians, generally used to hide information related to civil work projects that are privatising public spaces, and encouraged people to raise their voice, become active citizens and reclaim their own right to the city. I met Battah in February 2015, at the 'Resilient urban waterfront' international seminar organized by the LAU. I was a speaker together with Amira Solh, senior urban planner at Solidere. During the panel discussion, several questions from the audience, and from Battah, were on the general vision of Solidere, especially referring to the loss of publicness in the city centre. The BCD occupies only 10 per cent of the municipality extension, but it contains nearly half of the entire city's public open space. I argued, with a little provocation, that probably Lebanese people simply didn't feel the need for public space, or rather for that kind of public space that resembles Western urbanism, being profoundly different in terms of ownership, openness and democratic, social engagement. Solh argued that 'public space is a process of learned citizenship', something Lebanese people were asked to get used to again.

While the BCD is the most carefully planned district of the city, the suburban tissue is made of diverse, vibrant clusters, as a result of a progressive fragmentation of architecture and urban planning activities, delegated by the municipality to private entities and spontaneous dynamics (Maskineh, 2012). Since the end of the war, the attention to the city centre overshadowed most other urban areas. All around the glittering urban heart, the taste of the Arab culture beautifully spreads out in Achrafieh, Hamra, Gemmazye, Mar Mikahel and Sodeco, in a charming atmosphere, made of local shops, restaurants, flavours, sounds, symbols and humanity mixed together. But far away, in the southern suburb, Dahiya, where Hezbollah has a strong presence, terrorism and poverty are seriously affecting the population, made of multi-ethnic and religious groups, living in a constant state of emergency, while at the Shatila camp, with Palestinian and Syrian refugees[9] – a place that the United Nations Relief and Works Agency (UNRWA) describes as 'extremely bad' – there is a serious emergency in terms of environmental health conditions, including damp and overcrowded shelters: seawater is running from the taps, the electricity is cut every hour, gunfights occur regularly and 90 per cent of residents are unemployed.

As the sociologist Samir Khalaf argued (1993: 17–18), 'one of the most profound consequences of the war has been the redrawing of Lebanon's social geography'. Describing the disarray of the land-use patterns, often in absence of government and municipal authorities, and the decrease of mixed and heterogeneous communities, reinforcing distance and diffidence, Khalaf asks 'how is one to repair the fabrics of a fractured and dismembered society?'.

Lebanon: a multi-confessional society in search of the public good

For its geopolitical location, at the crossroad between East and West, Lebanon has served, throughout the centuries, as a rich environment for cultural exchange between different populations and communities. Renamed as the 'Switzerland of the Middle-East', for the significant importance in terms of politics and economics, and its ability to attract, during the 1960s and 1970s, Western finance and trade – also for its glamorous and luxury lifestyle – Lebanon was able to promote throughout its glorious history a unique experience of multiculturalism, where different confessions were intermingled but still recognizable, in an atmosphere of mutual understanding and cooperation.[10] The civil war completely devastated the country's economy, with a massive loss of human life. An estimated 150,000 people were killed, 184,000 injured; 13,000 kidnapped; and at least 17,000 missing. In addition, about 175 towns were partially or completely destroyed, and over 750,000 Lebanese were internally displaced. The physical damages in the country were estimated at $25 billion (Ghosn and Khoury, 2011: 382). Many left the country, creating a diaspora that moved millions of Lebanese residents to Europe, the United States, Canada, South America, Australia, and the Gulf. Recent years have seen the physical return of many from the diaspora, and others are investing in the reconstruction of Beirut, by owning shares in Solidere, or by taking part directly in the development of individual buildings.

In 1992, a new constitution was adopted to introduce into political life 'harmony between confessionalism and democracy' (Kheir, 2003), asking the Lebanese society to embrace diversity and tolerance. In principle, that was intended to be a common value for citizens to share, as it was in the past, but the devastating effects of the war generated separation and mistrust. It divided Lebanon into confessionally homogeneous geographical regions – like in Beirut, where the Green Line separated the Muslim West side from the Christian East side, without any possibility of crossing it during the whole period of the war except for those who had a specific authorization – and exasperating discourses about intercultural dialogue (Nahas, 2008).

Today Lebanon is a multi-confessional state, made of eighteen different religions: every citizen belongs to a confession, and this is part of the multiple Lebanese identity. The major groups active in the country are Shia (non-Orthodox), Sunni (Orthodox) and Druze Muslims, with Maronite, Greek Orthodox, and Greek Catholic Christians, and some Jews. There is also a significant Palestinian population, which has largely expanded since 1970, and an Armenian population since the genocide occurred in 1915.The religious system ensures civil rights, linking everyone to the political system, because political parties tend to be affiliated to a particular confessional group. The Taif agreement, signed in Saudi Arabia in 1989, which ended the civil war, rebalanced Lebanese politics, shifting power away from the Christian population, so that nowadays the parliamentary system guarantees the representation of the major confessional groups in the country: the President must be a Maronite Catholic Christian, the prime minister a Sunni Muslim, the deputy prime minister an Orthodox Christian and the speaker of the Parliament a Shia Muslim (Stewart, 2013). The portfolio of Cabinet members is also carefully chosen to reflect the confessional balance in the country as a whole.

After the assassination of the Prime Minister Rafik Hariri, on Valentine's Day in 2005, a period of political instability seriously influenced the civil society, putting a clear distance from the achievement of a real state of democracy. After 1990, Lebanon experienced waves of political and civil assassinations of prominent figures, like Samir Kassir, a journalist, an intellectual, social activist and influential columnist at the leading Lebanese daily *An-Nahar*, founding member of the Democratic Left movement, who was killed by a car bomb in Achrafieh (see Figure 14.2). Other events extended unrest and turmoil, between politics and society: random street skirmishes, a serious Israeli airstrike in 2006 that bombed the Beirut airport villages, army bases, bridges and a television station for one month (July), and, in November 2015, a pair of suicide bombings, in the southern suburb of Beirut, inhabited mostly by Shia Muslims, the worst episode since the end of the civil war, just a day before the terrorist attack in Paris at Bataclan. In 1997, during my first visit in Lebanon, the city of Sidon, where I was based, was bombed as well. These two episodes, in 1997 and 2015, that I personally experienced, reminded me, like a slap in the face, of the uncertainty and fragility of Lebanon and the scary threat of a history made of death and war, devastation and violence. Surprisingly, Beirutis consider these episodes a kind of norm, as part of 'a never entirely absent fear that another outbreak of violence may come'.[11]

The search for the public good is mainly related to a stable and lasting democracy, overcoming divisions, conflicts and interests of different confessions, which means freedom of expression, as an individual and as a community, exchange of cultural diversity, balancing also inequalities and rights. In his book, Kassir (2004) powerfully described the Arab's malaise, their misery,

Figure 14.2 A square in the BCD is entitled to the memory of Samir Kassir, a few steps
away from the Al-Omari Mosque, one of the biggest and oldest mosques in
Lebanon. Picture by the author.

wretchedness, unhappiness, based on a sense of victimhood and on the feeling
of being rejected by the Western world, questioning their identity in modern
society. The Arab's malaise is the result of objective realities in their countries
(Kassir, 2004; Kassab, 2009): the high illiteracy rate, the gap between rich and
poor, the overpopulated cities and deserted rural areas, the unresolved mourning
over a past greatness, but mostly the bitter awareness of helplessness in not being
in control of their own destiny.

Kassir argued that the globalization process through the use of the Internet
'could be Arab culture's great chance', several years before the Arab Spring and
its consequences.

Public space, media space and Arab urban hacking

In the Middle East, starting from 2010, several public spaces served as a means
for social activism and as a stage to deliver a powerful message, perceived as
highly courageous and strong in the Western world: in Avenue Habib Bourguiba
in downtown Tunis (2010–2011), a protest against the state of emergency due
to high unemployment, poor living conditions and oppressive political regime,
establishing what became the 'Arab Spring', a new era of civic boldness in the
Arab world, spreading across Egypt, Libya, Syria, Algeria, Jordan, Morocco and
several other countries; in Cairo's Tahrir Square (2011), leading to the resigna-
tion of the President Mubarak; also in the neighbour European-affiliated city
of Istanbul, in Taksim Square (2013), for the Gezi movement's struggle for the

preservation of the park as an urban common (that the local government wanted to replace with a shopping mall); and recently the You Stink protest in downtown Beirut (2015), using Riad al-Solh Square and Martyrs' Square as the main gathering places, was fighting to repossess the city, against rampant corruption of a privileged political class and governmental paralysis (the Republic of Lebanon experienced a 29-month political deadlock, since May 2014, for the vacancy of the President position, that was finally elected at the end of October 2016). These two squares were perfect locations for the rebellious to state their cause: the first one, next to the Grand Serail, the Governmental Palace, is the symbol of the authority they wanted to question, the second one is the place where the nation powerfully stood up against the Syrian army in 2005, right after the assassination of the Prime Minister Rafik Hariri, during mass demonstrations which were able to gather almost one million Lebanese citizens, thus ending a twenty-year presence that some had branded as an 'occupation'. Today, Martyrs' Square, from being a vibrant community space, even before the civil war, is a desolated huge, urban void, used mostly as a parking space, a vehicular intersection surrounded by building sites, with archaeological findings in the middle and the extremely huge neo-Ottoman mosque (Mohammad al-Amin Mosque, completed in 2007), standing on the West side, as the main heritage-like architecture (see Figure 14.3). This square is the most emblematic symbol of all conflicts and contradictions of the country, an emotionally moving site for public political expression.

Starting from a single space, as a symbolic powerful engagement, those protests expanded to multiple urban locations, incorporating more marginalized public spaces and publics, addressing long-standing socio-spatial inequalities between centre and periphery, attracting supporters and expanding civic wise. The protests

Figure 14.3 Martyrs' Square. Picture by the author.

worked also through a 'media space', mutually enforcing the relationship between physical places and social media, from a specific urban context and limited agenda to a more complex, huge and emotional human call in the public realm.

The Arab Social Media Report (ASMR), first published by the Dubai School of Government's Governance and Innovation Program in January 2011, showed that when it came to habits, more than half of the users use social media primarily to connect with people (55 per cent), revealing that chatting is the most common activity among users in the Arab world (50 per cent). The second report, published on May 2011, after the civil moments in Egypt and Tunisia, clearly showed that social media figured highly in both countries as a source of information, so that the number of Facebook users increased by 30 per cent in the first quarter of 2011.

The Arab cyberspace, in the MENA (Middle East and North Africa) region, served as a platform to exercise freedom of speech and as an easy gateway for civic engagement, using the media space to heavily reinvigorate the civil disobedience, mobilizing people and protesters with smartphones in hands, overcoming serious risks derived from clearly stating opposition to politics. Facebook and Twitter were used by activists, journalists and common people to play with specific hashtags, attracting thousands of likes and followers, exponentially increasing and spreading information about ongoing activities. Some of the commentators and journalists received aggressive messages by so-called Twitter trolls, trying to harass and intimidate them, especially when the tweet was highlighting the name of the country being discussed (Ghannam, 2012: 13).

'Hacktivism', the nonviolent use for political ends of 'illegal or legally ambiguous digital tools' (Hampson, 2012: 514), as the most devastating weapon of the digital age, has become a powerful means to generate actions of rebellion and protest against the consolidated system of power. Those actions shift from the virtual agora to the public physical realm, through different expressions of do-it-yourself urbanism (Iveson, 2013; Talen, 2015). Then, the place-based dynamics of real life are shifted back to the artificial world, keeping their distinguishing features, in terms of symbols, language and images, in order to be similarly recognizable in both contexts.

The TNS report presented at the Arab Social Media Influencers Summit (ASMIS) in 2015, found that of 7,000 social medial users in eighteen Arab countries, 55 per cent use social media to get news and learn new things. More than one third of the population in these countries is aged twenty-five or younger. This is significant since 67 per cent of people between fifteen and twenty-nine use Facebook. In the Arab world, WhatsApp is the most preferred channel (41 per cent). Smartphones are best friends in Arab everyday urbanism: information and news travel on instant communication platforms, being official and non-official at the same time, mixed with information coming from street life and personal beliefs. These freedom of speech and politically empowering activities, also taking place in Lebanon and especially in Beirut, are more and more able to establish a brand new idea of the city.

Some blogs, run by non-experts, are written in English, so that it is easy to follow the general discourse on Beirut's everyday life. Even if there is still a huge problem with regard to electricity and on an adequate broadband speed for the Internet connection, Lebanese people take huge advantage of the web, not only because they can much easily communicate in their own country, but mainly because they can place themselves on the spotlight in front of the world. From a Western perspective, for someone that never visited Lebanon, news and stories from the Middle East are delivered by newspapers and official media, that are somehow oriented: the idea of the Middle East is often related to the 'Arab world' in general, as a large, undefined geographical area, and to the assumption that it is affected by wars and conflicts, so that it is a dangerous place to visit and to live in. But a lot of counter sides of these news and stories are now available on local blogs, Facebook and Twitter profiles, provided mostly in English. The Lebanese are creating their own way to deliver a picture of the country, made of much more details and personal evaluations, using information coming from memories of old people, that are able to counterbalance the general image of desperation and poverty with a more nuanced and emotional point of view, giving voice to those neglected populations who are struggling every day for their own existence, thus providing also a platform for sharing and discussion at the local level.

Beirut's publicness and the right to the city: the role of artists

Besides the garbage crisis and the dysfunction of public services – such as electricity, water shortages, public transportation – that generated, in 2015, the You Stink movement, other large, bold protests and campaigns of rebellion, from local groups and open-minded citizens, are reclaiming public space, across the whole city of Beirut. Activists are reacting to extreme privatization trends that already eroded places for everyday public life and gravely put into risk the meaning of the city as a public good.

In 2012, Dictaphone Group, a multidisciplinary research collective, promoted the project *The Sea is Mine*,[12] to raise awareness about the privatization of the coastal line in the city of Beirut, to re-examine the understanding of public space and the right to use it, and to provide some occasions to reimagine the city. The daily Arabic language newspaper *Al-Akhbar* reported in 2012 that over 1,000 illegal resorts occupy the Lebanese coast,[13] which extends for almost 225 kilometres. Such legalized violation of the public land eroded the citizens' right to freely access beaches, in favour of luxury real estate projects, such as the Zaitunay Bay (see Figure 14.4), in the upscale marina built in BCD on a public property acquired by Solidere, designed by star architect Steven Holl, on the area that in ancient times was the fisherman port of the city. *The Sea is Mine* was conceived as a site-specific live performance, by Tania El Khoury, for ten days, exploring the accessibility of the sea, rules and laws that govern it, land ownership of the seafront and practices of its users. Participants were invited to take part in a journey on a fishing boat, from the Ein el-Mreisse port to Ramlet el-Baida beach, which is the last public beach in the city of Beirut (see Figure 14.8).

Several construction sites are currently reshaping Beirut's seafront: from Waterfront City in Dbayeh, along the highway towards Jounieh, a nearly $2 billion investment, to the Beirut New Waterfront District from Solidere, designed with

Figure 14.4 Saint George Hotel located between the Zaitunay Bay and The Corniche, the
seafront promenade. The hotel now stands as a relic, besides luxury real estate
developments, permanently showing an enormous banner with the words 'STOP
Solidere' on its bullet-scarred facade. Picture by the author.

the goal of expanding the BCD, to Ras Beirut, where different real estate developers are expanding existing private resorts. When the news about a new luxury project, by celebrity architect Rem Koolhaas, came out people were shocked: the Municipality was willing to sell to a private developer an officially protected natural reserve, the Dalieh beach in the Raouche area, a splendid natural landmark, further along the seaside promenade (Corniche), that was a popular swimming, diving and fishing spot and a grassy area for picnics for generations of Beirutis. Immediately the 'Civil Campaign to Protect Dalieh', as 'a coalition of individuals and non-governmental organizations who share a strong commitment to the preservation of Beirut's shared spaces, ecological and cultural diversity, as the pillars of the city's livability', was launched on Facebook in May 2014. This campaign engaged common people in weekly protests, university lectures and performances at Dalieh, also encouraging Facebook followers to share black and white photos of diving contests and family gatherings.[14] Activists wrote an open letter to Rem Koolhaas (December 2014),[15] promoted a crowdfunding campaign to support their activity, and an international ideas competition (March 2015) under the patronage of the Lebanese Ministry of Environment, in order to submit to the municipality three counter proposals – the winning projects were displayed at the Beirut Design Week (June 2015) – much more sensible to aspirations of Beirutis, to envision public use of the space. When the fence with razor wire was put in place by the police in 2014, to prevent the public access to the area, the creative group '5Ampere' put a banner on it, showing an old 10 liras banknote, with the image of Raouche from behind

Figure 14.5 The view of Raouche, also known as the Pigeon Rocks, from the Corniche, the seafront promenade, with the fence of razor wire dismantled by activists. Picture by the author.

a fence with barbed wire fence with the caption 'A country that is worth 10 liras'. Some artists elaborated creative paintings, such as *The Angry Dolosse Army* by Christian Zahr: pictures of this cartoon-like angry faces on breakwaters rocks, located besides Raouche, were posted by the artist on his personal Facebook profile.[16] Groups of artists started to work on the fence, creating human and animal figures with a plastic black tape, hanging drawings and signs. Waves of activists joyfully gathered with food and music, every week, gaining significant media coverage and momentum. After being up for a year, the 377-metre fence was torn down: tens of activists from the You Stink protest equipped with pliers pulled it down, opening the view again. Activists finally enjoyed the beach and repossessed the rocks and the sea. At the sunset, there was a sign replacing the fence saying 'The sea is ours' (see Figure 14.5).

The same courageous attitude guided some other activists, on September 2015 at Zaitunay Bay, that hosts luxury yachts and expensive restaurants. As a privately owned public space, people are not allowed to picnic, play music, ride bikes, rollerblade or other activities, as explained in a sign at the entrance, and the place is controlled by security guards. Activists organized a large picnic event and dance party, reclaiming the bay as a communal public space.

Publicness in Beirut is not easy to define, mostly because public space is quickly disappearing: the city seems to be driven by real estate developers, interested in perpetrating a massive substitution of traditional homes and businesses in favour of luxury towers, due to a zoning law continuously adapting for their needs, thanks to the complicity of government and authorities. The result is a serious impoverishment of the public urban space and playgrounds for children and a progressive lack of sensitivity about the public good, as a shared common value and a basic requirement for the society.

Save Beirut Heritage (SBH), an NGO committed to preservation of the architectural heritage, declared that the current number of remaining traditional homes and buildings in Beirut is close to 200, while at the beginning of the 1990s, a census counted around 1,600 of them. In the lack of specific planning rules, heritage owners, tempted by financial advantages suggested by developers, purposely allowed them to fall into disrepair, in order to remove them from the list of protected buildings. Sometimes through illegal demolitions and corruption, many ancient buildings were torn down, in a process affecting the whole city.

As one of the SBH activists claimed during the 2010 protest in Gemmayze, 'Beirut in time of peace has been much more disfigured and destroyed than in time of war' (Makdisi, 1997). SBH is also fighting to preserve some ghost buildings, that survived the war as ruins, like 'The Egg', as it's known, behind Martyrs' Square, an unfinished concrete structure, designed by the Lebanese architect Joseph Philippe Karam (1965), that was supposed to be the Beirut City Centre before the civil war (see Figure 14.6). Heritage activists enrich buildings with an emotional value and iconic meaning, as powerful echoes from the past, reminding of the devastation and the suffering from the war, as accidental monuments whose value is much higher than the economic profit.

While private interests are dominating public interests, a lot of graffiti around the city is giving voice to the social discontent. One of them, quite well-known, is based on a drawing by Jana Traboulsi, a Lebanese artist and graphic designer,

Figure 14.6 A construction site behind Martyrs' Square. On the left, 'The Egg'. Picture by the author.

depicting a skyline made of close towers surrounded by cranes with a caption in Arabic saying 'because they want to see the sea, we can no longer see the sky'. Other graffiti is free-hand Arabic texts, NGO logos, drawings from the urban and pop culture and street art (Zoghbi and Karl, 2011), standing together with over-sized billboards with faces of living or assassinated political leaders, showing across the city, and political messages, used by the military groups as a form of propaganda. 'White Wall Beirut', an event held in Beirut in 2012, celebrated the graffiti creative form of expression at the Beirut Art Centre and in many streets, with colourful murals of every scale, becoming the largest street art and graffiti exhibition in the Middle East.

One of the artists was Yazan Halwani, a young Beiruti, who also painted gigantic murals, in Hamra and Gemmayze, as large-scale portraits of Arab poets, musicians and actors, encircled by intricate Arabic calligraphy, including Samir Kassir, Khalil Gibran, the most known and read Lebanese poet, Fairouz, a Lebanese singer beloved all over the Arab world, and Sabah, a Lebanese singer and actress considered a 'Diva of Music' in the Arab World (see Figure 14.7). Through these prominent figures, Halwani tried to redefine the Lebanese culture, engaging communities around the walls before painting, to 'make sure the graffiti is growing with the city and not against it'.[17] People started to feel engaged with his beautiful murals and with the message they carry: 'the city can be changed and make it belonging to citizens instead of politicians or economic power, because citizens have a right to the city,' Halwani said.

Figure 14.7 Gigantic mural of Sabah in Hamra street by Yazan Halwani. Picture by the author.

All these art works, live performances and civic activism are filling the urban environment of new intangible meanings, playing with the imaginary dated back to the past, to those years made of wellbeing and prosperous growth. The sociologist Munir Khoury in the introduction of his book *What Is Wrong with Lebanon?* (1990: 7), before starting an exploration of the Arab world and its dilemmas, pluralisms and behaviours, stated that 'the good old Lebanon, that it is still living in the mind of millions of Lebanese at home and abroad, is gone forever; this is the price human beings, as individuals and as groups or nations, pay for ignorance and short-sightedness'. Khoury witnessed the destruction of the 'beautiful-ugly city of Beirut' and the disintegration of the identity of the civil society; nevertheless, in his book he strongly encourages Lebanese people to raise their voices and ask for a rebirth, a new '*Nahda*' (awakening). But many, even young generations, are disenchanted that anything could ever change in Lebanon, because of the leading power of private interests, the widespread corruption in the political environment and the powerful forces, also outside Lebanon, driving the local economy. As Kassir wrote (2004): 'Arabs are the unhappiest beings in the world, even when they don't realize it.' He strongly claimed that people of the Middle East should fight against the frustration of not being able to achieve independence, development or democracy, and take the future into their own hands.

On a mural next to the portrait of Samir Kassir, made by Halwani, under the Beirut River bridge in Karantina, the street artist Jean Kassir wrote: 'Frustration is not our destiny.' Fighting against this widespread sense of hopelessness, following an attempt to reclaim the city, moving from demanding to executing, at the

Figure 14.8 Ramlet El-Baida, the last public beach in the city of Beirut, in the southern end of the Corniche, the seafront promenade. Picture by the author.

municipal elections that took place in May 2016 in Beirut, a new grassroots movement called Beirut Madinati[18] (which means 'Beirut is my city'), a group made
of men and women from the civil society, of all ages and backgrounds, started to
work to address those issues that could let it become the place they know it can
be. Beirut Madinati, in a clear and well defined programme made of ten points,
answering ten major critical problems – such as urban mobility, greenery and
public space, housing, waste management, natural heritage, community spaces,
socio-economic development, environmental sustainability, health and safety,
governance – was giving hope to the people, was giving them the possibility of
having a voice and being heard. Even if they didn't win, they got up to 40 per cent
of the Beirutis' vote, thus demonstrating that a change is possible.

What I learned: a reflexive conclusion

The daily Beirut is a fulfilling experience made of intangible and sometimes unin-
telligible meanings, related to memories and symbols: there is a kind of creative
disorder under the chaotic urban surface made of unexpected human infrastructures, as a substantial layer of the Arab everyday urbanism, strictly related to art
and activism.

When the Sursock Museum in Achrafieh re-opened in the autumn of 2015,
after a seven-year renovation, the ground floor of the exhibition space was entirely
dedicated to the city of Beirut, described as 'a cacophony of cities within cities,
of disparate spaces existing simultaneously within and on top of one another'.
The exhibition, 'The City in the City', brought together works by artists, designers and researchers, such as action-film sets, mapping, photography, outcomes of
workshops and urban derives, to explore different spaces of contemporary Beirut,
enriching the debate on public space. In particular, the section called 'Practicing
the Public', a collective work developed by Ahmad Gharbieh, Mona Fawaz, Public
Works – Abir Saksouk-Sasso and Nadine Bekdache – and Mona Harb, was raising
questions on the meaning of 'public space' in Beirut, who used the space and how
such spaces are negotiated, appropriated and shared, through encounter, transactions and practices. Divided into four studies, 'Inhabiting the Street', 'Policing the
Street', 'Territorializing the Street' and 'Militarizing the Street' – this last one is
mapping visible security mechanisms, including checkpoints, barriers and army
personnel (Fawaz, 2009, 2012) – it gave a clear picture of the complexity of the
'public' dimension of the city of Beirut, which is not comparable at all with the
Western context.

As Sami Atallah wrote, on the Lebanese Centre for Policy Studies website,[19]
'in Lebanon the chief purpose of the three key political posts – the president,
the speaker of the parliament, and the prime minister – is to distribute resources
to different community leaders rather than serve people's needs and the public
interest.' This opinion is quite common among different social groups that seem
to be disenchanted from Lebanese political leaders, perceived as members of a
corrupted and money-oriented system. However, the political power seems not

to care about the consequences of a tangible, rising intolerance. A route towards political justice in Lebanon, after the assassination of Rafik Hariri in 2005, has become seriously complex and is being studied by international scholars and analysts, mainly from the West (Europe and the United States). This troubled country today has the power to influence the destiny of the entire world, because of the delicate relations with Iran, Syria, United Arab Emirates and Saudi Arabia, and the never-ending conflict with Israel. Lebanon is, geographically-speaking, the weakest country in a context of fractured lands (Anderson, 2016), after the invasion of Iraq thirteen years ago, leading to the rise of ISIS and the global refugee crisis. Even in such geo-political complexity, surrounded by wars and fighting parties, while individuals from eighteen different religious communities only agree to disagree, constantly reinforcing individualism, divisions and conflicts, a community-based movement supported by artists is boldly reclaiming the right to the city. As Harvey pointed out (2003), 'the right to the city is not merely a right to access what already exists, but a right to change it after our heart's desire'. While Beirut is shaped by the socio-economic, political and cultural context, imagination, memories and aspirations play their part: the urban environment embeds two different intertwining levels of interpretation, one tangible and the other intangible, both contributing to the production of space and its meanings. Public space in Beirut is neglected, abused, violated, destroyed but at the same time desired, imagined, reclaimed, appropriated. Rather than being an end product, Lebanese public space is a process (Mitchell, 2014) with multiple players, its narrative is fragmented in space and time, based on a reflexive process of gradual self-awareness (Lynch, 2000).

These findings are of great importance for an urban designer as I am: I'm supposed to design the urban space and to relate it to architecture, to create places where the community can gather, such as squares, streets, playgrounds, public facilities and resources. In one word, I'm supposed to design a city, or part of it, for people who live and inhabit it, so I need to understand the society and its urban culture, its structure and complexity. Otherwise I'm just superimposing a blind, empty, preconceived design to a context, whatsoever it is. What I learned in Beirut is that people can effectively impact on the urban environment: resilient citizens and artists are giving new shape to the Lebanese identity, reacting against distress, helplessness, victimhood and malaise. Beirutis are changing themselves to change their world, after decades of violence and suffering, they don't seem to be tired of fighting again. But it's their world, as it has been shaped, that changed them, if not all at least a part of them.

My understanding of the Arab urbanism followed the same route: as soon as I started to be aware of the complexity and richness of the hidden substratum of cultural Arab meanings, I realized that a different approach, deprived from many established Western (pre)conceptions, was inevitable, together with a brave sensitivity, open to embracing something mostly unknown, taking the risk to be hurt. And it did hurt, but I was already enchanted by the heart wrenching beauty and madness of Beirut.

Acknowledgments

I would like to acknowledge the Lebanese American University, who invited me to present and discuss some contents of this paper at the international seminar 'Architecture, Politics and Society' in February 2014, jointly organized by the Department of Architecture and Interior Design and the Department of Political Science and International Affairs. A special thanks also to the Lebanese University for the invitation to give a lecture, in December 2015, on my research findings on public space in the city of Beirut and for the fruitful discussion and feedbacks from scholars and students that followed. A grateful thanks to Robert Saliba, professor at the American University of Beirut, and to George Arbid, director of the Arab Centre for Architecture, for the generous interest in my research work and for providing useful insights and a relevant support.

Notes

1 According to the participant's briefing pack, edited by Dr Brian Hanson, Director of the Projects' Office of the Prince of Wales' Institute of Architecture, coordinator of the Urban Design Task Force in Lebanon.
2 According to the Lebanese Ministry of Education and Higher Education, www.higher-edu.gov.lb/english/default.htm [accessed August 2016].
3 Human Rights Watch (2015), Unequal and Unprotected. Women's Rights under Lebanese Personal Status Laws, www.hrw.org/report/2015/01/19/unequal-and-unprotected/womens-rights-under-lebanese-personal-status-laws [accessed August 2016].
4 In 2010, the most current annual per capita income data was US$17,090. In 2015 the GNI per capita (the gross national income converted to US dollars) of the Lebanese economy was between $4,036 and $12,475.
5 Sherwood, S. and Williams, G. (2009), 'The 44 places to go in 2009', *The New York Times*, www.nytimes.com/interactive/2009/01/11/travel/20090111_DESTINATIONS. html [accessed February 2016].
6 Wainwright, O. (2015), 'Is Beirut's glitzy downtown redevelopment all that it seems?', *The Guardian*, www.theguardian.com/cities/2015/jan/22/beirut-lebanon-glitzy-down-town-redevelopment-gucci-prada [accessed August 2016].
7 www.beirutreport.com/2014/01/erasing-memory-in-downtown-beirut.html [accessed February 2016].
8 www.youtube.com/watch?v=Bm2kDSqd5zo [accessed February 2016].
9 Fisher, A. (2015) 'This is daily life' in Shatila refugee camp', *Al Jazeera*, www.aljazeera. com/news/middleeast/2014/12/daily-life-shatila-refugee-camp-2014122863012886512. html [accessed February 2016].
10 According to Harb (2006), 'in political science terminology, confessionalism is a system of government that proportionally allocates political power among a country's communities—whether religious or ethnic—according to their percentage of the population'.
11 Barnard, A. (2015) 'Beirut, also the site of deadly attacks, feels forgotten', *The New York Times*, www.nytimes.com/2015/11/16/world/middleeast/beirut-lebanon-attacks-paris.html?_r=1 [accessed August 2016].
12 www.dictaphonegroup.com/work/the-sea-is-mine/ [accessed February 2016].
13 Zbeeb, M. (2012) 'Lebanon's seafront aggressors: The names and the details', *Al-Akhbar*, english.al-akhbar.com/node/14271 [accessed February 2016].
14 Battah, H. (2015) 'A city without a shore: Rem Koolhaas, Dalieh and the paving of Beirut's coast', *The Guardian*, www.theguardian.com/cities/2015/mar/17/rem-kool-haas-dalieh-beirut-shore-coast [accessed February 2016].

15 www.jadaliyya.com/pages/index/20264/open-letter-to-mr.-rem-koolhaas [accessed February 2016].
16 www.facebook.com/339041139521197/photos/?tab=album&album_id=643629 685729006 [accessed February 2016].
17 Bramley, E. V. (2015) 'How a Beirut graffiti artist is using his murals to try to unite a fragmented city', *The Guardian*, www.theguardian.com/cities/2015/sep/22/beirut-graffiti-artist-yazan-halwani-lebanese [accessed February 19, 2016].
18 beirutmadinati.com/?lang=en [accessed on April 2016].
19 www.lcps-lebanon.org/. Founded in 1989, the Lebanese Centre for Policy Studies is an independently managed, non-partisan, non-profit, non-governmental think tank whose mission is to produce and advocate policies that improve governance in Lebanon and the Arab region [accessed February 2016].

Reference

Anderson, S. (2016) *Fractured Lands: How the Arab world came apart*, The New York Times.

Bravo, L. (2017) 'Beirut in a frame. Layers of meanings on an ever-changing urbanity'. In *La Città Che si Rinnova*, Milano: Franco Angeli (in press).

Fawaz, M., Gharbieh, A. and Harb, M. (2009) *Beirut: Mapping Security*, Diwan series, Newspaper edition, International Architecture Biennale Rotterdam (IABR).

Fawaz, M., Harb, M. and Gharbieh, A. (2012) 'Living Beirut's Security Zones: An Investigation of the Modalities and Practice of Urban Security', *City & Society*, 24 (2): 173–195.

Flyvbjerg, B. (2006) 'Five Misunderstandings About Case-Study Research', *Qualitative Inquiry*, 12 (2): 219–245.

Gavin A. and Maluf R. (1996) *Beirut Reborn. The Restoration and Development of the Central District*, London: Academy Editions.

Ghosn, F., & Khoury, A. (2011). 'Lebanon after the civil war: Peace or the illusion of peace?', *Middle East Journal*, 65 (3), 381-397. DOI: 10.3751/65.3.12.

Ghannam, J. (2012) *Digital Media in the Arab World One Year After the Revolutions*, Center for International Media Assistance, National Endowment for Democracy.

Hampson, Noah C. N. (2012) 'Hacktivism: A New Breed of Protest in a Networked World', *Boston College International and Comparative Law Review*, 35 (2): 511–542.

Hanson, B. (1997) 'More than a memory'. *Perspectives*, December/January, 48–49.

Harb, I. (2006) *Lebanon's Confessionalism: Problems and Prospects*, The United States Institute of Peace, Washington, available at: https://www.usip.org/publications/2006/03/lebanons-confessionalism-problems-and-prospects [accessed March 2017].

Harvey, D. (2003) 'The right to the city'. *International Journal of Urban and Regional Research*, 27 (4): 939–941.

Iveson, K. (2013) 'Cities within the city: Do-it-yourself urbanism and the right to the city'. *International Journal of Urban and Regional Research*, 37 (3): 941–956.

Jones Hall, L. (2004) *Roman Berytus: Beirut in Late Antiquity*, London and New York: Routledge.

Kassab, E. S. (2009) *Contemporary Arab Thought: Cultural Critique in Comparative Perspective*, New York: Columbia University Press.

Kassir, S. (2004) *Considèrations sur le Malheur Arabe*, Paris: Acted Sud.

Khalaf, S. (1993) *Beirut Reclaimed: Reflections on Urban Design and the Restoration of Civility*, Beirut, Lebanon: Dar an-Nahar.

Kheir, W. (2003) 'Confessionalism and democracy: the case of Lebanon'. *God in Multicultural Society: Religion and Politics, Religion and Globalization*, Beirut,

Lebanon: Notre Dame University and Lebanese American University Joint Research and Value Philosophy Seminar.

Khoury, M. (1990) *What is Wrong with Lebanon? A Sociologist Examines the Lebanese Crisis*, Beirut, Lebanon: Al Hamra.

Lynch, M. (2000) 'Against reflectivity as an academic virtue and source of a privileged knowledge', *Theory, Culture & Society*, 17 (3): 26–54.

Makdisi, S. (1997) 'Laying Claim to Beirut: Urban Narrative and Spatial Identity in the Age of Solidere', *Critical Inquiry*, 23 (3): 660–705.

Mitchell, D. (2014) *The Right to the City: Social Justice and the Fight for Public Space*, New York: The Guilford Press.

Mashallah News and AMI (2014), *Beirut Re-Collected*, Lebanon: Tamyras.

Maskineh, C. (2012) 'Beirut, city of clusters: Planned city and urban chaos', *Area*, 120: 20–29.

Nahas, Georges N. (2008) 'The university experience, a way to meet the Other. A Lebanese Case'. Proceedings of the 4th International Barcelona Conference on Higher Education, Vol. 5. *The Role of Higher Education in Peace Building and Reconciliation Processes*, Barcelona: GUNI – Global University Network for Innovation. Available at www.gunirmies.net.

Robinson, J. (2011) 'Cities in a World of Cities: The Comparative Gesture', *International Journal of Urban and Regional Research*, 35 (1): 1–23.

Robinson, J. (2016) 'Comparative urbanism: New geographies and cultures of theorizing the urban'. *International Journal of Urban and Regional Research*, 40 (1): 187–199.

Sawalha, A. (2010) *Reconstructing Beirut: Memory and Space in a Postwar Arab City*, Austin, TX: The University of Texas Press.

Saliba, R. (1997) 'The Prince of Wales' Urban Design Task Force in Lebanon: the difficult reconciliation of Western concepts and local urban politics', *Urban Design International*, 2 (3): 155–168.

Seidman, S. (2011) 'From the Stranger to "the Other": The Politics of Cosmopolitan Beirut'. In McLaughlin, J., Phillimore, P. and Richardson, D. (eds.) *Contesting Recognition: Culture, Identity and Citizenship*, London: Palgrave Macmillan, 95–121

Stewart, D. J. (2013) *The Middle East Today: Political, geographical and cultural perspectives*, London and New York: Routledge.

Studio Beirut (2006) *Beyroutes: A guide to Beirut*, Archis, Amsterdam.

Talen, E. (2015) 'Do-it-yourself Urbanism: A history'. *Journal of Planning History*, 14 (2): 135–148.

Tuéni, N. (1979) *Liban: vingt poèmes pour un amour*, Beyrouth: Zakka.

Zoghbi, P. and Karl, D. (2011) *Arabic Graffiti*, Berlin, Germany: From Here to Fame Publishing.

15 Conclusion

Towards the worlding of art and the city?

Jason Luger

> How about art in *this* place – it is art to make people think again. Bring culture to your doorstep. We are not bringing madness. We are not bringing craziness.
>
> (Interview, Artist (Painter), M, Singapore, 2013)

From San Francisco

In 2015, the Chinese dissident artist Ai Wei Wei installed a site-specific piece at Alcatraz Prison in San Francisco. In the installation (called *@Large*), exiled, jailed and assassinated political activists from around the world were represented through mosaics, kites and paintings spread throughout an abandoned warehouse of the old prison. In this type of artwork, the art was locally-constituted, with Alcatraz Prison taking on particular meaning – but was also stripped of local context: Ai Wei Wei's Chinese-ness (in this case) was as site-irrelevant as the relationship between Alcatraz and Tibet or South Africa. Alcatraz likewise became a proxy representing all prisons, all those exiled or detained, moving beyond its site surrounded by the cold, churning waters of the San Francisco Bay. The San Francisco skyline, glittering and tantalizingly just out of reach to the inmates once imprisoned in the island – came to represent all cities. The artist – and artwork – were both trans-national and stateless; local and global both represented in place, as international tourists walked around the site taking photographs (many of which were then beamed to social media, perhaps then re-Tweeted thousands of miles away). This tableau at Alcatraz embodies the contradictions, inter-connections, disjunctions, symphonies and dissonances of the critical global art-scape, and its relationship to place.

Contemplating at this juncture the relationship between art and the city, I cannot help but approach this topic through the lens of my San Francisco home. At once, all the paradoxes and contradictions of art and its relationship to the city are visible: the uneasy coexistence between political street art in the Mission district and the rapid gentrification underway; the dispersal of working artists to far flung suburbs while the elite of (wealthy) Pacific Heights and the tech economy fill their homes with priceless artistic treasures; the use and co-option of, but the simultaneous liberation and empowerment of the digital art-scape by the hometown tech giants like Google, Facebook and Twitter. Art cannot be divorced from the hyper-capitalist tech city, nor can the city from the art which helped make it famous.

Indeed, looking back in local history, it was the art provocations of the 1960s – the 'Beat Poets' howling at the moon from North Beach; the street performers and street provocateurs (such as the 'Diggers') in the Haight-Ashbury during the 1967 'Summer of Love'; the mass-participatory public art encounters such as the *Human Be-In* (in Golden Gate Park, 1967) – that helped give rise to the somewhat anarchic, destructive, brilliant and creative forces of Silicon Valley today. Art creates the urban realm; art helps to change, and sometimes, to destroy it. Art and the city are constantly being made, and re-made, operating within and helping to shape the currents of power at various scales, fixed in territory and moving across space and time. Art and the city are, have always been, and will always be, inextricably linked. Artists have always gravitated toward urban themes, help- ing to shape the urban conversation, whether in utopian forms such as Constant Nieuwenhuys's *New Babylon* (1959), or dystopian nightmares such as Picasso's *Guernica* (1937). We owe our conception of cities to artists.

Nor can art be divorced, ever, from urban politics and larger social and cul- tural changes underway. As Simon and Garfunkel sang (1964), 'the words of the prophets are written on the subway walls' (in the form of street-art); Hitler was radicalized as a fin-de-siècle Vienna artist. As this volume has shown, art is cen- tral to the relationship between urban space and authoritarianism; democracy; the market; and the activist, whether in Lebanon or London; Los Angeles or Queens; Hong Kong or Hackney; Singapore or Sweden. When gentrification or failed urban redevelopment projects tear neighbourhoods and human souls apart, it is often art that remains to tell the story – representations of memories, of dreams, of hope. Emblems of false promises or symbols of triumph. Whimsy and sadness, joy and despair, sometimes at the same time. Are these not also inherently urban emotions? Wirth (1938) recognized the emotional tapestry that urban life pre- sented; Debord (1956) and later Rancière (2009) understood how art and art-led provocations are central to understanding, and experiencing, urban life. Deutsche (1998) continued the discussion on art, the city and spatiality, critically suggesting that art, urban space and policy have combined in ways convenient to the neolib- eral agenda of displacement, and *eviction*. In her framing, art and the city do not necessarily have a relationship conducive to social justice but rather to reinforce uneven power relations and urban development agendas. It is questions like these, explored by Deutsche, that deserve further interrogation, and which this volume seeks to address in a paradigm in which neoliberalism – and its associated produc- tion of urban spaces – has circulated globally for decades and remains dominant, even if dead or dying (Smith, 2008).

Thus, I conclude this volume by calling for a much-needed joining of art the- ory and urban theory. This volume has been one such attempt to engage with, explore, expose and critique the global urban art-scape, through the use of case studies in diverse terrains. Chapters have dealt explicitly with the way that critical art transforms the place-making process; the way that art transforms institutional structures and re-arranges hierarchies; the way that art gives voice to those normally without. In this concluding chapter, I will invoke some of the key themes and issues emerging from within this volume in order to ask some lingering questions, in the hope that researchers can continue to take forward the important task of mapping the contours of art in the city. But first, I will address some of the lingering problems plaguing urban theory, which are now by default also problems plaguing urban art-theory, since art is urban, and *the urban, in all its forms, is art.*

The lingering questions surrounding *art and the city*

There is no single global urban theory, despite decades of attempts to form one by urban theorists. Developing a unified urban theory has proven difficult, given the breadth and diversity of places, trends, actors and networks: 'Cities are notoriously fuzzy at the edges, variegated within and differentiated from place to place. This is one good reason why coming up with a "theory of the urban" has never been an easy matter' (Walker, 2016: 177). There remain a number of un-answered questions in urban theory, and these questions can be extended to art, and its relationship to the city. This volume has engaged with several of these questions, helping to push the discussion forward. Notably, three main questions have emerged: One is the way that scale and systems require exploration, in terms of how one place, one artist or one artwork relates to another; the second is the way that urban identity – and by extension, artistic urban identity – is intersectional, constantly in flux, rather than fixed. The third is the way that 'place' relates to immaterial space – whether that be the digital realm or global thought/idea circuits; and within that, the way that 'local' intersects, sometimes awkwardly, with transnationalism and globalism. These discussions lead to several ripe areas for a research agenda going forward, where this chapter concludes.

Scale and systems

Scale remains a problematic concept in urban theory. Authors such as Walker (2015, 2016) and Scott and Storper (2014) have reasserted the importance of both the local scale, and the relationship between actors, networks, institutions and flows at a variety of relational scales. This return to scale comes after others (Brenner, 2013) have suggested that place no longer matters, and that all urban discussion should take place at the global level. Cities, and within them, artists, and indeed artworks, are often rooted in multiple scales at once. Walker (2016: 174) differentiates between scales into 'absolute, relative, and relational space', in terms of the linkages between urban neighbourhoods, systems, and networks: 'Urban areas need to be repeatedly taken apart and put back together again in order to see how they work, even in terms of how intra-urban networks, creative encounters ... operate' (Walker, 2016: 175). The art-scape exists within and across all these scales and likewise deserves to be taken apart, explored, and put back together again.

Authors have also struggled to define and categorise the ways that systems of cities link together in a comparative frame (and at different scales). Some (Scott and Storper, 2014) seek to do away with regional or categorical distinctions such as 'North' and 'South', 'East,' or 'West' toward a global urban theory; others (Roy, 2009; Roy and Ong, 2011; Robinson, 2006, 2012) argue for the importance of such distinctions in building theory. This volume engages with this debate with chapters suggesting that these geographical/socio-cultural and political distinctions are crucial in understanding the way that cities, networks and artists are linked between one place and another. In art, in particular, setting is crucial: the same artwork carries different meaning if performed or observed in Germany than it does in Hong Kong (as Kam, Chapter 8 demonstrated), or in Kabul, Johannesburg, or Nairobi. Pink paint (Alibhai/Arboleda, Chapter 1) in Johannesburg comes to signify homelessness; but pink can just as easily signify breast cancer, gay rights or nothing at all. Meanwhile, Jacobi (Chapter 6) brings to light the way that art schools – one 'grouping' of global institutions, are scaled up and down within the context of the United Kingdom's increasingly neoliberal higher education establishment; a back room at a pub can serve as a proxy for art education when a formal institution becomes inaccessible.

Intersections/identity

Exploring the intersections of identity and belonging are trendy in recent literature and theory; an acceptance that identity is multifaceted and not fixed corresponds with a broader paradigmatic shift where assumed stable categories and delineations are being reconsidered as geographical theory seeks to move beyond post-colonial notions of state, city and society. 'Place' has been re-imagined as relational rather than static (Massey, 1995), as have the set of state-society relations (Jessop, 2001). *Relationality* can also be extended to individual identity formation/negotiation, and the construction of meaning, intention and agency. This has necessitated new, broader, definitions of *activism* and *the activist*, which are found at both local and global scales, bound to territory and de-territorialised; art is likewise locally rooted and global in scope. *Intersectionality* has become a useful theoretical framework at a time when race, class, gender, sexuality, culture and politics emerge and combine in complicated and sometimes fraught ways, in different global settings.

Sometimes conspicuously absent from the emerging discussion on art/ activism are explorations of the diversity among, and between, artists. The arts are often un-interrupted from the elite networks, spaces and topics they inhabit. Authors such as Markusen (2006) have probed the class differentials and varying aims/agendas of cultural producers, but this literature pales in comparison to more plentiful investigations of elite-driven cultural policy, cultural producers and cultural spaces. The ways in which art-making, class, race, sexuality, policy and space intersect are crucial to understand the way that artistic identity is forged and the way that art should and can be interpreted and portrayed. Methodological and ethical realities likely reinforce these elite-focused studies. Therefore, *intersectionality* may allow a more inclusive approach to theory, which can also be extended to activism's many forms and aims, and the way that art and activism intersect. Keenly aware of these intersections, Sonia-Wallace (Chapter 7) exposes his own attempts to bring the sometimes ironic interplays (and sometimes awkward, yet necessary self-reflection) between marginalized cultural producer and elite cultural consumer through his 'Rent Poet' performances in Los Angeles.

Art theorist Sharon Irish (Irish and Lacy 2010: 11) asserted that, 'the majority of artists [including women artists] are not social activists'. If this is true, then where are the intersections between art and activism? Where, in the post-modern, digital age, can and should theorists draw the lines between art and politics, while keeping in mind their own positionality and the ethical implications of their (privileged) gaze as an observer? These questions become particularly compelling moving away from North American/European settings, often the focus of such investigations (see Bravo, Chapter 14, who reflects upon her role as an outside observer in the context of Beirut's art/activism). *Just being an artist* with no deliberate intention to provoke may nonetheless provoke; attempts to produce political art may not resonate; and an artist's self-identity is in flux, not stable. Parry (Chapter 9) found himself at this junction, caught between the 'Climate Games' whimsy and the darker subtext of France's 'state of emergency', in which activities not normally associated with anti-state activism were fenced off and made 'other', in the context of a climate of paranoia and Islamophobia.

Why should an artist working in one place have an automatic solidarity with an artist working somewhere else? Or with the oppressed, in a general sense? Popke (2007) surveyed the consensus amongst humanist and post-structural urban theorists such as Amin (2004: 43) who claimed that 'cities are locked into a multitude of relational networks of varying geographical reach'; this idea built upon Massey's (1995) earlier conception of a 'global sense of place'. Therefore, artists – particularly those urban artists in highly-connected global cities such as Singapore – are likewise relationally connected not only to other global artists but to global citizenry and the idea of what Madden (2012) called 'city becoming world'. If cities become 'world', then by extension, the artist becomes world. This frame can also be applied to the artists' situating her/himself within global activism: Facebook activism (or 'click-tivism') as novel and endearing as it may seem, is still governed by the Silicon Valley corporation. Even softer forms of global activism may come as uninvited guests – does, and must, 'Black Lives Matter' have automatic resonance with socially-engaged artists everywhere? The answer again depends on artists' self-identity and the under-explored, multifaceted nature of art-activist agency and autonomy. The way in which the critical artist herself deals with mundane, everyday issues and engages with (or stems from) working class culture represents another gap in theory; the disconnections between the *global artist* and the working-class artist brings to mind earlier explorations of 'cultural capital'/'habitus' (Bourdieu, 1986) but deserves further contextual exploration. Milic (Chapter 12) engages with this topic, exploring the intersection of class, urban space, gentrification and art through her provocation with the Southwark community. She grapples with the sometimes uncomfortable tension of both being a displaced cultural producer herself and producing work which represents the displacement of the broader working class, posing complicated questions for how to theorize class/art/space in the city.

Kester – and other art theorists – have called for a re-examination of the way the interaction between artist, audience and activism is interpreted and represented. Kester (in Kucor and Leung, 2012: 157) drew upon Habermas's concept of identity formed through a constantly shifting discursive process, one in which the artist should not be 'conceptualized as actualising his or her will through the heroic transformation of nature or the assimilation of cultural difference,' but rather, understood to be in a constant state of negotiation, 'one defined in terms of open-ness, of listening and a willingness to accept dependence and intersubjective vulnerability' [ibid.: 157]. Therefore, the identity of the artist her or himself, and the meaning of the work created performed, is formed in the intersection between self and other, or as Kester asserted, 'in the interstices between the artist and the collaborator' [ibid.: 157]. Hedemyr (Chapter 11) showed us how these interstices can be sharply brought into focus when the artist uses the urban environment in unexpected ways; a policewoman interrogates a 'suspicious' activity in a city park and becomes part of the artist's re-working of the urban environment unknowingly. Trees, even, become actors in the artist-audience-environment relationship, as Hedemyr portrayed in the park in Gothenburg.

The material/immaterial (urban) artscape; tensions between 'local' and 'global'

Returning to Ai Wei Wei's installation in San Francisco, the tension between place and no place, local and global presents itself. Ai Wei Wei, in representing worldwide dissidents in an Alcatraz warehouse, has rendered himself stateless, existing in a sort of grey-area of global artistic themes. *@Large* is a site-specific artwork only in the sense that the (physical) space was large enough for the installation. It could have just as easily been staged almost anywhere in the world, perhaps less sensationally, and without the views of the sun setting over the Golden Gate. Coombs (Chapter 4) touches upon this theme, overviewing the satirical 'Ghana ThinkTank': site-specific provocations rooted to one place yet at the same time stateless, global and stripped of local meaning. In her exploration, Coombs hints at the amorphous and disconnected nature of global think tanks and NGOs (whether the World Bank or well-meaning philanthropic foundations) and their sometimes-uneasy relationship to the local communities they seek to serve. Teo and Tien (Chapter 2) also grapple with the tension between place (in their case, a fixed gallery space in Singapore) and no-place (a more nomadic existence, free of a postal address) by highlighting the crucial importance – and role played – of 'Post Museum' both for Singapore's art-activist community as well as the local immigrant community. A material node,

'Post Museum' allowed for transformations, and emancipations, not often observed in space, place, and ideology – restricted Singapore.

Hawkins (2013: 53) suggested that theory should correspond to art's 'expanding field' in which artistic practice is no longer limited to the canvas or to material space at all. In particular, Hawkins noted that 'socially engaged art' increasingly takes new forms, ranging from 'gallery-based cook offs' to the mobilization of artistic protest (ibid.: 56). Hawkins also noted the that there still 'lingers a very modern conception of art as a 'separate sphere' of objects located primarily in the gallery of public square' (ibid.: 60), rather than a true theoretical connection between artistic practice and the variegated landscapes of political movements. These connections, Hawkins stressed, require further understanding, particularly, 'how to practice and assess the interdisciplinary scholarship that sees art being used as an empirical source for geographical study' (ibid.: 66). Correspondingly, art and its networks remain firmly rooted to place, and local context is crucial to understanding art/policy assemblages. Landau (Chapter 5) demonstrates this by her exploration of the 'composition and constellation' of Berlin's art-policy field, composed of unique actors at specific moments.

Where amongst the intersection of art and activism are the boundaries? What are the intentions, and how should readings of meaning change from place to place? As Kester (2012) summarized, 'increasingly, avant-garde art sought to challenge, rather than corroborate, conventional systems of meaning' (in Kucor and Leung, 2012: 155). Some of these twentieth-century movements included the Situationism/Dada-ism; surrealism; and Cubism movements; and the various ways the 1960s gave rise to art-activism in different places (such as the Haight-Ashbury, San Francisco, that I mentioned previously). Yet this perceived 'rupture' (Kester, 2012: 155) of the avant-garde invites revisiting in the current paradigm of reinvigorated global activism: the Europe of Dada's provocations cannot serve as a proxy for all contexts in understanding the relationship between art, politics, and the interpretation of artistic intention and meaning; especially as online platforms from YouTube to Pinterest now play an integral role in the *artscape*, cutting across formerly static boundaries (such as national borders).

Indeed, as cases from Beirut (Bravo, Chapter 14), Hong Kong (Lu and Wong and S. Wong, Chapters 3 and 13), Taipei (Lu and Wong, Chapter 3) and Singapore (Teo and Tien, Chapter 2) demonstrate – social media and local context combine in ways not observable, or replicable, elsewhere. And yet simultaneously, movements such as #YouStink (Beirut) are inextricably global, instantaneously joined by hashtags and satellites. Maeder *et al.* (Chapter 10) even question whether earlier conceptions of the subversive and transformative power – and critical potential – of public art in Western Europe are even still appropriate, given current trends and changes to institutional and policy apparatuses. These compelling cases are worthy contributions to the ongoing discussion on art and the city, but perhaps their biggest contribution – and that of this volume – is to bring into focus some clear questions that researchers are invited to explore further in the quest to develop, and enrich, urban theory.

Towards the worlding of critical art and the city?
(Questions for a research agenda)

> All good art is political! There is none that isn't.
> (Toni Morrison, 2008)

This volume is only the beginning of what promises to be fecund terrain for researchers going forward in the continuing task of theoretically constructing *art and the city,* and many questions remain deserving of more exploration. Some of these are uncomfortable: Can the artist be authoritarian? Can the artist be complicit with those they critique? Must art make us think, challenge our values, inspire, provoke? To what degree do we – as observers – inspire, motivate, or challenge the artist? Is it time to re-think the often unchallenged association of the art-scape with social justice, with an assumed critical alliance with other just causes?

In what way is the art-scape deeply rooted in place; contextually variegated; contingent on micro interactions and micro-places? Correspondingly, if the urban is now planetary, stretched across the world infinitely and instantly, then is art also planetary, caught up in the amorphous web of global flows? Is there such a thing as local art anymore? Within this web, what are the sites and nodes of intra-artist relations, occurring at different scales, relationally?

Just as there are city-systems or groupings of cities, are there also global groupings of artists, systems of art, that can or should be categorized and defined? What, and where, are the differentiations between elite and non-elite artists and art-forms, often circulating in close proximity but sometimes separated by vast chasms of class and power? Can art, and artists, ever be separated from the elite? Has it, and they, ever been?

Can the artscape ever be separated from, or explored independently of, global capitalism and the circulation of capital? If so, what would this sort of exploration look like – given the realities of institutional hierarchies, funding, and the need for both researchers and artists to increasingly demonstrate impact and value?

What of time and temporal processes? Some elements of the artscape are instant; some occur over a short period; some are displayed for some set amount of days, week, years; some are permanent. How to go about developing theory when the researcher is often limited to fixed sites, at fixed periods of time? When one snippet of time neither encapsulates the true scope and duration of an artwork, nor the social movements or political cycles the art often responds to, then how can the researcher build a meaningful case?

Finally, is it even possible to develop a 'global' theory of *art and the city* – does *worlding* this conversation really necessitate appreciating the ways that art and its relation to cities is so different, that the art-scape may be incomparable? *To what do we compare the artscape?* If a truly cosmopolitan, locally constituted, contextually diverse art-urban theory is to develop, then it must be decolonized from the elite centres of both art and theory; the periphery must be given the platform it requires to gain centrality. This extends not only to the periphery of global power relations, but also to the periphery of academic disciplines: neither art nor the city are 'owned' by any one ivory tower. This volume was one such attempt, but further efforts must go much farther. Art and the city have never been more important, or imperilled. The critical artscape is worthy of such explorations.

References

Amin, A. (2004) 'Regions unbound: towards a new politics of place'. *Geografiska Annaler, Series B, Human Geography* 86: 33–44.

Bourdieu, P. (1986) 'The Forms of Capital'. In J. Richardson (ed.) *Handbook of Theory and Research for the Sociology of Education*, New York: Greenwood, 241–258.

Brenner, N. (2013) 'Theses on urbanization'. *Public Culture* 25(1) 85–114.

Debord, G. (1956) 'Theory of the "Derive"'. *Situationist International Online*. Translated by Ken Knabb. [Retrieved 2016-07-12.]

Deutsche, R. (1998) *Evictions: Art and Spatial Politics*. Cambridge, MA: MIT Press.

Hawkins, H. (2013) 'Geography and art, and expanding field: Site, the body, and practice'. *Progress in Human Geography* 37(1), 52–71.

Irish, S. and Lacy, S., (2010) *Suzanne Lacy: Spaces Between*. Minneapolis: U of Minnesota Press.

Jessop, B. (2001) 'Institutional (re)turns and the strategic-relational approach'. *Environment and Planning A* 33(7), 1213–35.

Kester, G. (2012) 'Conversation pieces: The role of dialogue in socially engaged art'. In Kocur, Z. and Leong, S. (eds.) *Theory in Contemporary Art Since 1985*, Oxford, UK: Wiley Blackwell, 153–165.

Madden, D. J. (2012) 'City becoming world: Nancy, Lefebvre, and the global–urban imagination'. *Environment and Planning D: Society and Space* 30, 772–787.

Markusen, A. (2006) 'Urban development and the politics of a creative class: evidence from a study of artists', *Environment and Planning A* 38(10), 1921–1940.

Massey, D. (1995) *Space, Place and Gender*. Minneapolis, MI: University of Minnesota Press.

McFarlane, C. and Robinson, J., (2012) Introduction—experiments in comparative urbanism. Urban Geography, 33(6), pp.765–773.

Morrison, T. (2008) *Poets and Writers Magazine*, Nov/Dec 2008, story by Kevin Nance.

Popke, J. (2007) 'Geography and ethics: spaces of cosmopolitan responsibility'. *Progress in Human Geography* 31(4), 509–518.

Rancière, J. (2009) *The Emancipated Spectator*, London, UK: Verso.

Robinson, J., (2006) *Ordinary Cities: Between Modernity and Development* (Vol. 4). London: Psychology Press.

Roy, A. (2009) The 21st-century metropolis: new geographies of theory. Regional Studies, 43(6), pp.819-830.

Roy, A. and Ong, A. eds., (2011) Worlding Cities: Asian Experiments and the Art of Being Global (Vol. 42). London: John Wiley & Sons.

Scott, A. and M. Storper (2014) 'The nature of cities: the scope and limits of urban theory'. *International Journal of Urban and Regional Research* 39(1), 1–15.

Smith, N. (2008) 'Comment: Neo-liberalism: dominant, but dead'. *Focaal* (51), 155.

Till, K. (2008) 'Artistic and activist memory-work: Approaching place-based practice'. *Memory Studies* 1(1), 99–113.

Walker, R. (2015) 'Building a better theory of the urban: A response to "Towards a new epistemology of the urban?"' *City* 19(2/3), 183–91.

Walker, R. (2016) 'Why do cities exist?' *International Journal of Urban and Regional Research*, published online, May 2016.

Wirth, L. (1938) 'Urbanism as a way of life'. *American Journal of Sociology* (44), 1–24.

Index

Page references in *italics* refer to figures and in **bold** refer to tables.

Printed and bound by CPI Group (UK) Ltd, Croydon, CR0 4YY

21/10/2024

01777055-0011